GRADED FERROELECTRICS, TRANSPACITORS AND TRANSPONENTS

T0189424

Multifunctional Thin Film Series

Editors:

Orlando Auciello, Argonne National Laboratory,
Ramamoorthy Ramesh, University of California at Berkeley

The basic applied science and applications to micro- and nano-devices of multifunctional thin films span several fast-evolving interdisciplinary fields of research and technological development worldwide. A major driving force for the extensive research being performed in many universities, industrial and national laboratories is the promise of applications to a new generation of advanced micro- and nano-devices. These applications can potentially revolutionize current technologies and /or create new ones while simultaneously creating multibillion dollar markets.) Multifunctional thin films cover a wide range of materials from metals to insulators to organics, including novel nanostructured metals, semiconductors, oxides, widebandgap, polymers, self-assembled organic layers and many other materials.

Series titles:

Nanoscale Phenomena in Ferroelectric Thin Films
Seungbum Hong
Graded Ferroelectrics, Transpacitors, and Transponents
Joseph V. Mantese and S. Pamir Alpay

GRADED FERROELECTRICS, TRANSPACITORS AND TRANSPONENTS

Joseph V. Mantese

S. Pamir Alpay

 Springer

Library of Congress Cataloging-in-Publication Data

A C.I.P. Catalogue record for this book is available
from the Library of Congress.

ISBN 978-1-4419-3605-9   e-ISBN 978-0-387-23320-8   Printed on acid-free paper.

Printed in the United States of America.

9 8 7 6 5 4 3 2 1

springeronline.com

Contents

Preface

It has been more than 80 years since Valasek first recognized the existence of a dielectric analogue to ferromagnetism, ferroelectricity, in Rochelle salt. Much as with semiconductor research, the initial studies of ferroelectric materials focused on homogeneous materials. Unlike semiconductor research, however, which rapidly expanded into non-homogeneous structures and devices, investigations of compositionally graded and layered ferroelectrics have been relatively recent endeavors. Indeed, many of the most significant results and analysis pertaining to polarization-graded ferroelectrics have only appeared in publication within the last ten years. Further extensions of these concepts to the general class of order-parameter graded ferroic materials, as depicted on the cover of this book, have (with one exception) been totally lacking.

It was thus with a great deal of excitement that we assembled the manuscript for this book. The primary focus of this study is directed toward polarization-graded ferroelectrics and their active components, *transpacitors*; however, the findings presented here are quite general. The theory of graded ferroics is put on a solid foundation in chapters 2 and 5; whereas, much of the introductory material relies more heavily upon analogy. This was done so as to provide the reader with an intuitive approach to graded ferroics, thereby enabling them to see heterogeneous ferroics as clearly logical extensions of passive semiconductor junction devices such as *p-n* and *n-p* diodes and their active manifestations, transistors, to: *transpacitors*, *transductors*, *translastics*, and ultimately to the general active ferroic elements, *transponents*.

There are many individuals who have assisted us in the preparation of this book and of whose work we have relied heavily upon in assembling our manuscript. Norman S. Schubring and Adolph L. Micheli, two of the discoverers of the phenomena of polarization-graded ferroelectrics, have provided enumerable hours of discussion, detailing both their experimental setups and their material preparation techniques. Greg W. Auner and Ratna Naik of Wayne State University have been our ever-constant experimental collaborators, bringing an infectious enthusiasm to the materials study. We have benefited greatly from many of the papers of Dinghua Bao, Hai-Xia Cao, Zen-Ya Li, and Carlos A. Paz de Araujo who have duplicated the original findings obtained from compositionally graded barium strontium titanate and extended the analysis to a variety of new material systems. Zhigang Ban's thesis work at the University of Connecticut (UConn) laid much of the foundation for the theoretical study of graded ferroics, his contributions being invaluable in placing the analysis of such structures on a solid theoretical footing. While chapters 2 and 5 are not definitive, they present the understanding of graded ferroics to date, and a roadmap for future theoretical efforts. We also are very grateful to Alexander L. Roytburd (University of Maryland) for our many discussions with him. He too has played a major role in the theoretical analysis and we have quite frequently sought his advice and wise counsel. In particular, we have tremendously benefited from the many discussions pertaining to the stability of the ferroelectric phase in bilayer structures.

Nearly all the fundamental research papers on graded ferroics first appeared in *Applied Physics Letters*. We thus wish to thank the journal editors and their staff for their foresight in supporting this area of research, especially Orlando Auciello who shepherded many of the early manuscripts on this subject through to publication.

This book certainly would not have been possible without the assistance of Gürsel Akçay, who read the many versions of the manuscript in their entirety and assisted us throughout its preparation. It was he who designed the cover of the book.

One of us (JVM) would like to thank Delphi, and Linos Jacovides in particular, for supporting the study and understanding of graded ferroics; and the other (SPA) is extremely grateful to the National Science Foundation (Division of Materials Research) who funded him and his students through a CAREER Grant. SPA would also like to thank his graduate students, Burç Mısırlıoğlu, Anuj Sharma, and Shan Zhong, Chris Gatto for proof-reading the manuscript, the Dean of School of Engineering of UConn, the Director of the Institute of Materials Science, and his colleagues in the Department of Materials Science and Engineering.

Writing this book was not the easiest of tasks, especially as we undertook it as an endeavor to be completed in our spare time. With the demanding schedules we both faced, much personal time with our families was sacrificed in order to complete the task in a timely manner. We both would like to thank our families for their patience, understanding and support. We dedicate this book to Anita, Paul, John, and David Mantese and Nesli and Ata Alpay.

And of course, this work benefited greatly from the assistance of Greg Franklin and Carol Day at Springer-Verlag, who led us through the publication process, patiently worked with us to correct the many mechanical deficiencies of our earlier drafts, and assisted us in finalizing the manuscript.

Delphi Research Laboratories   J. V. Mantese
Shelby Township, MI
Department of Materials Science and Engineering   S. P. Alpay
University of Connecticut, Storrs, CT
November 2004

Chapter 1

POLARIZATION-GRADED FERROELECTRICS: THE DIELECTRIC ANALOGUES OF SEMICONDUCTOR JUNCTION DEVICES

1.   INTRODUCTION

The invention and subsequent development of the transistor is arguably one of the most important innovations of the twentieth century. It is often the benchmark by which other technologies are measured and is the example most often cited as to what characterizes a significant break-through. Indeed, it is hard to overstate the impact of transistor technology on everyday life. Key to the creation of transistors and other active semiconductor components and devices, was the development of a means for breaking the spatial symmetry of the local transport properties found in the homogeneous base materials from which they are formed. As a consequence of chemical doping, local inhomogeneities in free charge density arise which result in built-in potentials that give diodes, transistors, and other active semiconductor elements their unique properties.

Given the tremendous importance of transistor technology, it is worthwhile asking if its analogue can be identified and created in material systems. It is thus from this perspective that we review recent efforts in the search for the dielectric analogue of the transistor, i.e., trans-capacitive (transpacitive) devices and structures. We find, however, that the analysis and concepts conceived for polarization-graded ferroelectrics have quite general validity, applicable to a wide range of other ferroic components and systems. Thus, we are able to expand our analysis to include graded ferroics and their active counterparts, transponents.

Our discussion of graded ferroics begins with polarization-graded ferroelectrics (also commonly referred to as graded ferroelectric devices, GFDs) that have a brief, yet rich, history. Indeed, it was soon realized by many of the authors whose work we cover in this book, that they were creating conceptually new devices and structures along the lines of a transistor analogue. Hence, much of the material in this book follows a historical account, as the research on polarization-graded ferroelectrics was often conducted as a means of confirming the obvious connection to chemically graded semiconductor devices. Following the approach used by the principal researchers in the study of graded ferroics (particularly ferroelectrics) we reason by analogy throughout much of this book.

We begin our discussion of polarization-graded ferroelectrics, and transpacitors with a cursory review of semiconductors and the fundamentals of diode junction devices. Like our predecessors, we draw analogies between the variations in local free charge density found within *p-n* junctions and the spatial dependence of the local dipole density found in chemically non-homogeneous ferroelectrics. It will be seen that in direct analogy with non-homogeneous semiconductors, a local potential will dominate the electrical response of graded ferroelectrics.

2.   REVIEW OF SIMPLE SEMICONDUCTOR STRUCTURES

This short review of homogeneous semiconductors and junction devices is not meant to be comprehensive, but serves only to present the salient features and physics of homogenous semiconductors, diodes, and transistors, and to provide a basis for comparison with GFDs. We shall not specifically discuss the band structure of semiconductors; however, it is responsible for the local free charge density, the mobility, and a host of other electro-optical and physical properties of the materials as well as their temperature dependences. There are, of course, a great many excellent texts and review articles which can instruct the interested reader as to the fundamentals of both diode junction devices and more complicated semiconductor structures (Ashcroft and Mermin 1976; Streetman 1980; Sze 1981).

A compositionally uniform bar of semiconductor material (aligned along the *z*-axis in simple three dimensional space) having a uniform free charge density, $n(z)=n_0$, yields a linear (symmetric) current-voltage characteristic (*I-V* plot) when non-rectifying contacts are attached to its ends, see Figure 1.1. Strictly speaking, the local charge density in the vicinity of the contacts is non-uniform. However, we assume for this discussion that the doping near the contacts is sufficiently large that we can neglect such effects. The conductivity, $\sigma=n\mu q$ (with μ the free charge mobility and q the charge of the carriers), is proportional to the slope of the material's *I-V* plot, with the proportionality constant a function of the sample geometry. For such materials, the fundamental quantities of interest are charge flow (current) and the driving potential (voltage).

Of course, it is not chemically homogeneous semiconductor materials and structures that have spawned the tremendous growth in the electronics industry, but their non-homogenous counterparts, specifically diode junctions and transistors. A basic understanding of the response of most semiconductor devices begins with *p-n* or *n-p* junctions; for it is from an understanding of these simple structures that we are able to grasp the workings of more complex devices.

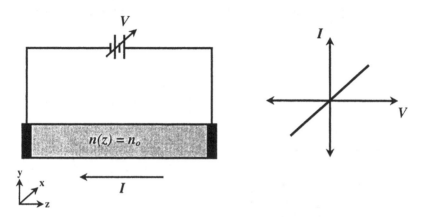

Figure 1.1. Compositionally uniform semiconductor aligned along the z-axis, with constant free charge density, $n(\mathbf{r}) = n_o$, displaying symmetric (linear) *I-V* characteristics.

From semiconducting materials, diode junctions may be formed which have asymmetric current-voltage characteristics. The asymmetry in these junction devices arises from the non-uniform, local doping of the semiconductor to produce regions where the number and type of local charge carrier are made to be prescribed functions of position, $n=n(\mathbf{r})$, where $\mathbf{r}$ is a directional vector. Specifically, a *p-n* junction device is formed when *p*-type (doped with an electron acceptor) and *n*-type (doped with an electron donor) semiconductor materials are placed in contact with one-another and affixed with non-rectifying contacts, see Figure 1.2. Free charge inter-diffuses across the contact region leaving behind uncompensated, fixed, ionic charge densities (the space charge region) that gives rise to a position-dependent potential $V=V(\mathbf{r})$, which suppresses further charge flow across a plane interface aligned normal to the z-axis in the absence of an external field. In Figure 1.2, and Equation 1.1 below, we assume equal and opposite free charge densities across the interface for the sake of simplicity.

The potential may be obtained by solving Poisson's equation on either side of the junction:

$$\nabla^2 V(\mathbf{r}) = \pm \frac{e \cdot n(\mathbf{r})}{\varepsilon} \qquad (1.1)$$

The negative sign is taken in the *n*-doped region ($z \geq 0$) and the positive sign for the *p*-doped region ($z \leq 0$). Boundary conditions are matched at the interface. Here, ε is the dielectric constant of the host semiconductor material, and *e* is the electronic charge (magnitude). We also have assumed, for simplicity, an abrupt junction with large carrier density supplied by singularly charge dopants.

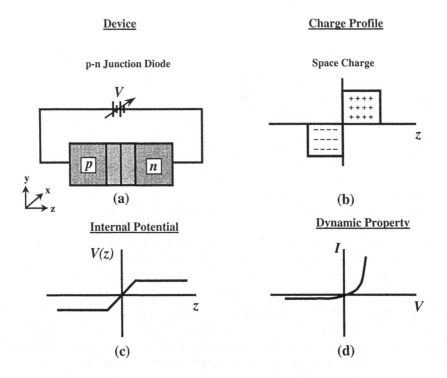

Figure 1.2. (a) *p-n* diode device. (b) The space charge region at an abrupt interface. (c) The internal potential, $V(\mathbf{r})$, as a function of position. (d) The asymmetric *I-V* characteristics of the device. We again align the semiconductor along the *z*-axis as discussed in the body of the text.

For variations of $n(\mathbf{r})$ only along the *z*-axis, the above relation reduces to:

$$\frac{d^2 V(z)}{dz^2} = \pm \frac{e \cdot n(z)}{\varepsilon} \qquad (1.2)$$

It is apparent from an integration of Equation 1.2 to yield $V(z)$, that a "built in" potential, ΔV, can be associated directly with the junction; corresponding to the potential difference across the semiconductor depletion region, such that $\Delta V=V(L)-V(0)$, where L is the width of the depletion zone. For a symmetrically doped abrupt junction, ΔV is logarithmically related to the density of free charge carrier density in the junction region (Streetman 1980). It is the existence of ΔV which gives rise to the asymmetries observed in junction devices. It is also important to note that ΔV exists even in the absence of an externally applied field (though it cannot be measured directly because of the presence of the free charge), and is the manifestation of the local material chemistry.

In the semiconducting field, "forward" and "reversed" biased junction devices yield asymmetric *I-V* characteristics, dependent upon the sense of the applied voltage and whether one has present a *p-n* or *n-p* junction, see again Figure 1.2. Under forward bias, ΔV is opposed by the application of an external field, and the junction device readily conducts, thereby allowing free charge to transition across the space charge region. Under reverse bias, the blocking potential, ΔV, is further augmented and the flow of free charge is inhibited, resulting in a structure that cannot support free charge flow. Clearly, from Figure 1.2, a *p-n* junction device may be spatially inverted to yield an *n-p* structure.

The spatial and electrical asymmetries are what distinguish semiconducting junction devices from their homogeneous counterparts. It is, therefore, from this point of understanding that we make the leap to dielectric media and graded ferroelectric structures.

3.   SIMPLE GRADED FERROELECTRIC DEVICES

The dielectric analogue of a semiconductor must be a material that has an electric dipole density whose magnitude and direction are intrinsic to the material and which may be tailored to yield local regions of directed polarization, $\mathbf{P}=\mathbf{P}(\mathbf{r})$. For simplicity of discussion, we assume that the ferroelectric material can be described by a polarization intensity directed along the z-axis as might be for the case of tetragonal perovskite ferroelectrics; where the one dimensional spatial variations in the polarization may be expressed as, $\mathbf{P}=[0,0,P(z)]$.

In the discussion that follows, we only consider structures wherein the local dipole asymmetry is a function of the internal material chemistry, analogous to our previous discussion of semiconductor junction devices. We do not consider induced asymmetries that might arise as a result of non-homogeneous applied potentials; although later in this Chapter we will extend our analysis to local variations in dipole density that arise due to temperature and stress non-uniformities. Such materials constitute the class commonly referred to as ferroelectrics. For dielectrics, specifically the sub-class of ferroelectrics, the relevant quantities of interest are charge (Q), and voltage (V). The interested reader can find many comprehensive reviews on the subject of ferroelectrics (Jona and Shirane 1962; Lines and Glass 1977).

An understanding of the dynamic properties of ferroelectrics may be gained from an examination of the thermodynamic potentials that govern such materials (Ban, Alpay et al. 2003). Once again, however; we introduce in this overview chapter only a superficial thermodynamic analysis and reserve the complete theoretical treatment for Chapter 2 where we take into account electrostatic and electromechanical interactions due to spatial variations in the polarization. For a homogenous ferroelectric element, unconstrained by any clamping forces, the free energy may be expressed (approximately) in terms of the polarization, P, as:

$$F(P) = F_0 + \frac{1}{2}\alpha P^2 + \frac{1}{4}\beta P^4 + \frac{1}{6}\gamma P^6 \qquad (1.3)$$

Where F_0 is the energy in the paraelectric state, α, β, and γ are expansion coefficients which are functions of composition and temperature; and are such that, below a critical temperature, T_C (referred to as the Curie temperature) a double-well potential arises that accounts for the two spontaneous polarization states, $\pm P_S$, of the material, see Figure 1.3. Note that $F(P)$ is only a function of the magnitude of P, not its direction. This aspect will be of importance when we later consider polarization-graded devices.

The polarization of a ferroelectric material for temperatures below T_C is spontaneous, non-zero, and present even in the absence of an applied external field. Thus, the local polarization density $P = P(z)$ will be seen to be analogous to the local free charge density $n = n(z)$ found in doped semiconductors.

The electric field E may then be solved for as:

$$\frac{dF}{dP} = E = \alpha P + \beta P^3 + \gamma P^5 \qquad (1.4)$$

which must vanish for $P = \pm P_S$, corresponding to the spontaneous polarization of a homogeneous material, again see Figure 1.3.

A ferroelectric of uniform composition will yield a symmetric hysteresis loop (Q-V plot, or equivalently a P-E plot when Q and V are normalized to the dimensions of the material) for capacitor structures formed from the material.

Upon application of a periodic potential, V, the individual dipoles in the ferroelectric device fall into alignment, producing an internal field that opposes the applied potential. The total polarization of the structure may be switched between its two polarization states with the polarization reversals occurring at critical electric fields. The switchable polarization of the device saturates to a finite value having positive or negative orientation as shown in Figure 1.4.

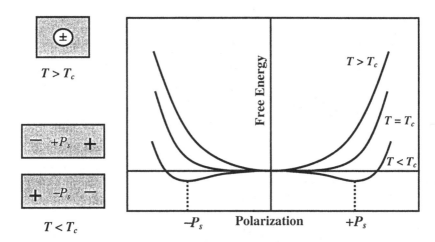

Figure 1.3. Free-energy diagram of a homogeneous ferroelectric at a temperature below its critical (Curie) temperature, showing two possible states for its polarization, $\pm P_s$ that yield minima in the energy.

To discuss non-homogenous ferroelectrics (with the hope of discovering the dielectric analogue of the *p-n* junction) we must discover a means for creating an internal potential within a simple ferroelectric that is intrinsic to its structure. However, unlike semiconductor junction devices whose potentials arise from a diffusion of free charge across chemically graded junctions, the intrinsic potentials in graded ferroelectrics must be constructed from gradients in bound charge or dipole moment density (Mantese, Schubring et al. 1995).

The local dipole density of a ferroelectric, **P**, is a function of the material chemistry and both the external field and any depoling field that might arise due to the orientation of the internal dipoles. Much as with our previous discussion of semiconductor materials, the polarization density can be made to be a function of the local position by suitably tailoring the material chemistry. Hence, it is possible to construct structures such that $P=P(z)$ (Mantese, Schubring et al. 1995).

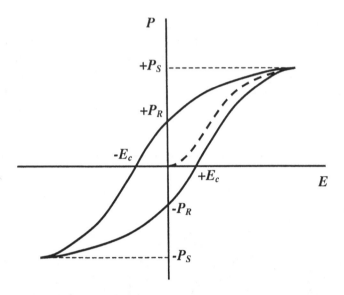

Figure 1.4. A typical polarization versus applied electric field hysteresis curve of a ferroelectric, displaying symmetric *P-E* characteristics. P_S: Spontaneous polarization, P_R: Remanent polarization, E_C: Coercive field.

For the case where we only consider vector variations along the z-axis the dielectric displacement, $D=D(z)$, is related to the polarization density via the relation (Jackson 1998):

$$D = \varepsilon_0 E + P \qquad (1.5)$$

where ε_0 is the permittivity of free space. In the absence of free charges, $dD/dz=0$, whence we can relate the local potential $V=V(z)$ to the polarization via the expression:

$$\frac{d^2V(z)}{dz^2} = \frac{1}{\varepsilon_0}\frac{dP}{dz} \qquad (1.6)$$

which is the dielectric counterpart of Equation 1.2 for semiconductors. Note, however, in Equation 1.2, the free charge density was assumed to be constant in both the p and n doped semiconductors with $n(z)$ discontinuous only along the plane $z=0$; whereas, $P(z)$ was assumed to vary continuously in the discussion above.

Note also, that V is not a simple function of P, for as depicted in its related *P-E* plot, see Figure 1.4; the potential is a non-linear, multi-valued function of P. Furthermore, Equation 1.5 contains contributions to V due to both an induced dipole moment (sometimes referred to as the permittivity or by the misnomer – dielectric constant) and that due to the spontaneous polarization, P_S, of the material. Finally, P itself is a function of V through the local electric field, $E= E(z)$. We shall assume henceforth that our ferroelectric materials are perfect insulators so as to make the analysis straightforward, although we will relax this restriction in Chapter 2.

To discover how an internal potential, ΔV, arises within a non-homogeneous ferroelectric in the absence of any applied potential, we therefore, return to Equations 1.3 and 1.4. For a homogeneous ferroelectric, Equation 1.4 vanishes due to the fact that the free energy, as expressed by Equation 1.3, is a constant function of position for constant polarization with $P=P_S$. The material can then be described thermodynamically by sets of double well potentials of constant depth and width arrayed along the z-axis as shown in Figure 1.5a. However, suppose P were to vary linearly with position along z, then:

$$\frac{dP}{dz}=\frac{P(z=L)-P(z=0)}{L}\equiv\frac{\Delta P}{L}=const., \qquad (1.7)$$

where L is the length of the ferroelectric along which there is a spatial variation in the polarization.

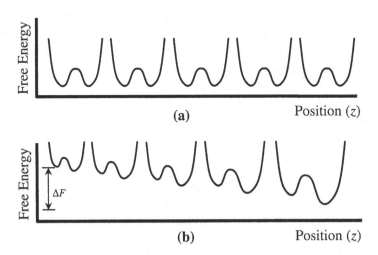

Figure 1.5. (a) Gibbs free energy diagram for a homogeneous material aligned along the z-axis. (b) Free-energy diagram of the graded ferroelectric structure. ΔF is the "built in" energy difference across the (linearly) polarization- graded ferroelectric.

From Equation 1.3 and 1.4 we thus have:

$$\Delta V = -\int_0^L E(z) \cdot dz = -\int_0^L \frac{\partial F(z)}{\partial P(z)} \cdot dz$$

$$= -\left(\frac{dP}{dz}\right)^{-1} \int_0^L \frac{\partial F(z)}{\partial P(z)} \cdot \frac{dP}{dz} \cdot dz$$

(1.8)

or,

$$\Delta V = -\left(\frac{dP}{dz}\right)^{-1} [F(P(z=L)) - F(P(z=0))] = -\frac{L}{\Delta P}\Delta F .$$

(1.9)

Equation 1.9 implies that, even in the absence of an external field, an internal potential is present within a polarization-graded ferroelectric. Consequently, one would anticipate that the free energy could be described as a series of skewed double well potentials as depicted in Figure 1.5b. Indeed, Vanderbilt and Rabe have shown from *ab initio* calculations that variations in local chemical environment break the symmetry of the ferroelectric perovskite system, giving rise to just the asymmetric potentials described above (Sai, Meyer et al. 2000; Sai, Rabe et al. 2002). Note, however, that the role of 180° domains needs to be considered for a complete description of polarization-graded ferroelectrics, something not fully taken into account in the theoretical literature as of this writing.

In analogy with our discussion concerning semiconductor junction devices, the intrinsic internal potentials of polarization-graded ferroelectrics must govern their overall dynamic response. Upon excitation of a graded ferroelectric with an externally applied periodic potential, several things happen. At very low fields, the ions rattle about their respective wells; there are no transitions to or from any of the randomly populated double wells. As the magnitude of the periodic applied potential is increased, the ions in the deepest wells still do not contribute to any hysteresis and remain locked in their double well sites at either the upper or lower well (but randomly occupied); however, there is now switching in the shallowest double well pairs. But the switching is not completely symmetric, for the lowest energy wells will have a higher probability of occupancy than the upper wells due to their lower energy. Hence, one now observes hysteresis, but this time with a displacement of the hysteresis graph along the polarization axis, reflecting some net population of the lowest wells. The presence of a net occupancy in the lowest wells shifts the potential imparted by the compositional gradient, causing all the wells to become even more asymmetric. At higher potentials, there is now switching of some of the deepest wells. The spontaneous polarization increases; now there are an even greater number of ions locked into the lowest double well sites. Further cascading and tilting of the wells continues as the applied potential is increased.

Analogous to semiconductor diode junctions that yield asymmetric *I-V* characteristics as a result of "built in" potentials, polarization-graded ferroelectrics exhibit asymmetric *Q-V* plots. Such structures have been shown to produce shifted charge-voltage hysteresis ("up" and "down") loops as depicted schematically in Figure 1.6 (Schubring, Mantese et al. 1992; Mantese, Schubring et al. 1995; Mantese, Schubring et al. 1997). The direction and magnitude of the shifts have been found to depend upon the sense of the chemical gradient (i.e., the degree and direction of polarization gradient) (Mantese, Schubring et al. 1997; Brazier, McElfresh et al. 1998; Tsurumi, Miyasou et al. 1998; Chen, Arita et al. 1999; Bao, Mizutani et al. 2000; Bao, Mizutani et al. 2000; Boerasu, Pintilie et al. 2000; Bao, Wakiya et al. 2001). Of course, as in the case of *p-n* and *n-p* junctions, an "up" device may be converted into a "down" device by a simple spatial inversion.

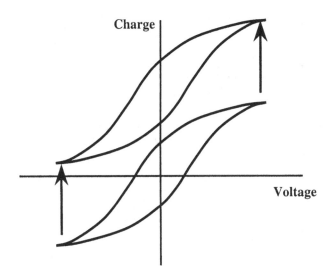

Figure 1.6. Schematic of the dynamic response (*Q-V* plots) of an "up" device. The terminology will become clearer in Chapter 3.

A gradient in polarization for a ferroelectric material may also be achieved by other methods; but those of particular interest include: (1) The imposition of a temperature gradient across the sample (Mantese, Schubring et al. 1995; Fellberg, Mantese et al. 2001), (2) A compositional gradient normal to the growth surface (Mantese, Schubring et al. 1997; Brazier, McElfresh et al. 1998; Tsurumi, Miyasou et al. 1998; Chen, Arita et al. 1999; Bao, Yao et al. 2000; Bao, Zhang et al. 2000), and (3), a stress/strain gradient across the sample (Mendiola, Calzada et al. 1998; Kim, Oh et al. 1999; Canedy, Li et al. 2000). Figure 1.7, in particular, illustrates the typical temperature dependence of a ferroelectric material's spontaneous polarization near its Curie temperature. Indeed, much as with semiconductors, true measurements of the properties of homogeneous materials are often masked by the presence of unwanted internal potentials that lead to spurious results. Figure 1.8 illustrates how polarization-graded ferroelectrics may be achieved through temperature, stress/strain, and compositional gradients.

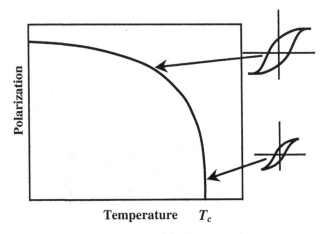

Figure 1.7. Plot showing the typical variation of a ferroelectric material's spontaneous polarization with temperature near its Curie point.

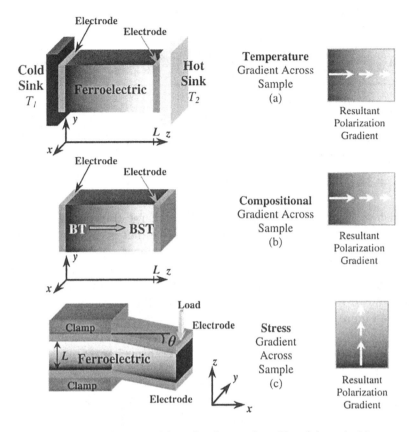

Figure 1.8. Polarization-graded ferroelectrics may be achieved through: (a) temperature, (b) stress, and (c) compositional gradients.

4. FURTHER ANALOGIES BETWEEN SEMICONDUCTOR JUNCTION DEVICES AND POLARIZATION-GRADED FERROELECTRICS

Table 1.1 below compares semiconductor junction devices with their graded ferroelectric counterparts. Semiconductors are charge flow devices (I, current) and thus lead to trans-resistive, "transistor" devices. Graded ferroelectric devices function by charge storage (Q, charge) and thus lead to trans-capacitive devices whose function is to control their charge storage capacity. We return to this discussion in Chapter 4.

Table 1.1. Semiconductors vs. Ferroelectric Devices

System	Density Function	Origin of the Internal Potential	Asymmetric Response	External Power Source
Semiconductor	$n(\mathbf{r})$, free charge density	$\nabla \cdot n(\mathbf{r})$	I - V	External DC supply voltage
Ferroelectric	$\mathbf{P}(\mathbf{r})$, dipole density	$\nabla \cdot \mathbf{P}(\mathbf{r})$	Q - V	External AC supply voltage

Transistors function by modulating the free charge transport through an *n-p-n* or *p-n-p* junction transistor to create a device that is capable of power gain. In semiconductor transistors, power gain is achieved when a small signal current is injected into the base region of a transistor and is amplified sufficiently to drive a load by means of transistor action (Horowitz and Hill 1980). The power sources for transistors are the direct current (*dc*) supply voltages, see Figure 1.9.

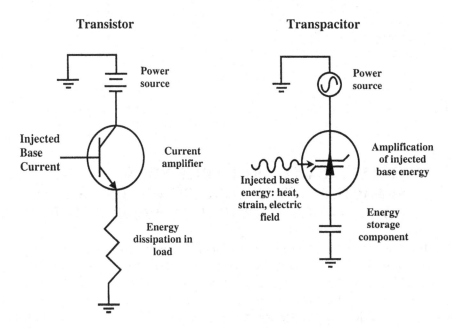

Figure 1.9. Comparison between a transistor and a trans-capacitive device, both configured for signal power gain.

Trans-capacitive ferroelectric devices are charge storage/voltage generating devices usually driven by an external alternating potential (*ac*) analogous to the transistor *dc* supply. In these latter structures, the internal potential (see Equation 1.9), ΔV, of a GFD is altered by a modulating energy flux to the device. The source of energy flux may take a number of forms including: heat, strain energy, or applied field. ΔV is also a function of the peak excitation voltage of the *ac* power source (Mantese, Schubring et al. 1997; Brazier, McElfresh et al. 1998; Schubring, Mantese et al. 1999).

5.   GRADED FERROICS: THE GENERAL CASE

Before closing this chapter, we remark that the above analysis can be made quite general. There exist a host of ferroic systems that can be characterized by free energy expressions similar to Equation 1.3, namely (Wadhawan 2000):

$$F = F_0 + \frac{1}{2}\alpha\eta^2 + \frac{1}{4}\beta\eta^4 + \frac{1}{6}\gamma\eta^6 \qquad (1.10)$$

where again, α, β, and γ are expansion coefficients which may be functions of composition and temperature, and η is the system-specific order parameter (Ban, Alpay et al. 2003). A general expression of the type written in Equation 1.10 may be used to describe not only ferroelectrics, but to ferromagnets, ferroelastics, ferrogyrotropic materials, and other unconventional systems such as high-T_C and BCS superconductors, superfluids, materials near critical points, and even giant magnetoresitive materials (Wadhawan 2000).

In the thermodynamic analysis of most ferroic systems, the $\nabla\cdot\eta$ term (and higher powers of it) is usually assumed to be small away from the phase transformation temperature, representing minor thermal fluctuations about an equilibrium value. In Chapter 2, we show for ferroelectric systems that this term is important if there are artificial gradients due to systematic spatial polarization variations. Hence the $\nabla\cdot\eta$ term becomes of primary interest, driving the unique properties of the systems. In Chapter 5 we consider other ferroic systems where $\nabla\cdot\eta$ serves to break the local spatial symmetry, permitting us to extend our analysis to graded ferroics; a general class of materials and devices.

Chapter 2

THERMODYNAMIC THEORY OF POLARIZATION-GRADED FERROELECTRICS

1.   INTRODUCTION

As discussed briefly in the previous chapter, the basis for the unusual properties of polarization-graded ferroelectrics lies in their "built-in" potentials. In this Chapter, a thermodynamic analysis is developed to explain the origin of the properties of polarization-graded ferroelectrics, most notably the charge offset, i.e., the displacement of the polarization vs. applied electric field along the polarization axis. We build on the preliminary analysis of Chapter 1, incorporating the electrical and electromechanical coupling into the basic free energy functional (Equation 1.3).

Theoretical modeling of polarization-graded ferroelectrics has a rather brief history. There have, however, been significant efforts in the analysis of ferroelectric superlattices and multilayer ferroelectric heterostructures from which an understanding of graded ferroelectrics may be built. The theoretical description of superlattices has gained significant momentum in the past years, mainly due to advances in thin film synthesis that allow exceptional compositional control and structural stability on the nanoscale (Erbil, Kim et al. 1996; Kanno, Hayashi et al. 1996; Kim, Gerhardt et al. 1997; Qu, Evstigneev et al. 1998; Specht, Christen et al. 1998; Jiang, Pan et al. 1999; Marrec, Farhi et al. 2000; Nakagawara, Shimuta et al. 2000; O'Neill, Bowman et al. 2000; Jayadevan and Tseng 2002; Shimuta, Nakagawara et al. 2002; Koebernik, Haessler et al. 2004) and, thus permit comparison to theory. We note that the modeling of multiple linear dielectric layers coupled to each other is not new and is covered in standard textbooks on dielectric materials (Anderson 1964; Jackson 1998; Scaife 1998). However, ferroelectrics, which possess a spontaneous polarization in the absence of an applied electric field, are not linear dielectric materials because of the strong non-linear electric field dependence of the dielectric permittivity. The modeling of multiple ferroelectric layers in intimate contact is thus a complex and challenging problem that must take into account interlayer electrostatic and electromechanical coupling. There are a variety of methods that have been employed to describe ferroelectric multilayers. Transverse Ising models

(TIM), phenomenological approaches based on the Landau formalism, as well as first-principles calculations have been devised in order to understand the underlying physics of such heterostructures (Cottam, Tilley et al. 1984; Schwenk, Fishman et al. 1990; Wang and Mills 1992; Li, Eastman et al. 1997; Qu, Zhong et al. 1997; Marvan and Fousek 1998; Ma, Shen et al. 2000; Shen and Ma 2000; Sepliarsky, Phillpot et al. 2001; Shen and Ma 2001; Wang and Tilley 2001; Bratkovsky and Levanyuk 2002; Ong, Osman et al. 2002; Chew, Ishibashi et al. 2003; Neaton and Rabe 2003). It is not our intention to review these numerous studies in detail, as this would lead us beyond the scope of this book. We will make use of some of the results and ask the reader's indulgence for omitting studies that may be relevant to the subject matter discussed in this Chapter.

One of the first attempts to explain polarization-graded ferroelectrics their behavior was based on a Slater model where the free energy was expanded in terms of the systematic spatial variation of the ionic displacements (Mantese, Schubring et al. 1997). There have also been two studies that employed TIM models to calculate the polarization profile in temperature-graded, compositionally uniform ferroelectric materials (Wang, Wang et al. 2001; Cao and Li 2003). Temperature- and compositionally graded ferroelectrics have also been analyzed using basic electrostatic considerations (Brazier and McElfresh 1999; Pintilie, Boerasu et al. 2003; Chan, Lam et al. 2004). Recently, a thermodynamic model was constructed based on the Landau-Ginzburg formalism, which took into account electrostatic and elastic interactions that provided a unified approach for three types of polarization-graded systems (Ban, Alpay et al. 2003). The model was shown to be in good quantitative agreement with published experimental results (Alpay, Ban et al. 2003).

The theoretical approach developed in this Chapter to understand the nature of polarization-graded ferroelectrics is based on the Landau theory of phase transformations. It was first introduced in 1930s to describe a complex problem in solid-state phase transformations-ordering phenomena in metallic alloys (Landau and Lifshitz 1980). The order-disorder transformation involves a change in the crystal symmetry. The material transforms from a high-symmetry disordered phase to a low-symmetry ordered phase (Guttman 1956; de Fontaine 1979; Ziman 1979). The broken symmetry in the crystal upon ordering can be characterized by an order parameter. Landau has shown that the (Helmholz) free energy of an order-disorder transformation can be expressed very simply as a polynomial expansion of the order parameter that describes the degree of order (or disorder).

Originally, the Taylor expansion of the order parameter was intended to be limited to temperatures close to the phase transformation temperature.

However, it was shown later that the polynomial form of the free energy functional provides a very good approximation even at temperatures far below the transition temperature should higher-order expansion terms be included in the free energy expression. Furthermore, the original theory was constructed for 2nd-order phase transformations, which do not show a discontinuity in the entropy, i.e., no abrupt change in the slope of the free energy versus temperature curve at the transition temperature. The heat capacity (as well as the elastic moduli), however, displays a λ-type anomaly in the vicinity of the phase transformation temperature. The Landau methodology can also be extended to include 1st-order phase transformation for which the entropy is discontinuous. It should be noted that the "jump" in the entropy must necessarily correspond to a commensurate change in internal energy, defined as the latent heat of the phase transformation.

Again, it is not our intention to provide details of the Landau formalism. For this, the reader is referred to textbooks, most notably to Landau and Lifshitz's *Statistical Physics* (Landau and Lifshitz 1980) or Salje's *Phase Transformations in Ferroelastic and Co-elastic Crystals* (Salje 1990). Landau theory has been used quite successfully to describe phase transformations in a variety of materials systems, including (but not limited to) ferromagnetic, martensitic (ferroelastic), and ferroelectric transitions with appropriate order parameters. For example, proper ferroelectric, ferromagnetic, and ferroelastic phase transformations can be described via the Landau potential with the polarization P_i, magnetization M_i, or the self-strain as the order parameter, respectively (Wadhawan 2000).

The extension of Landau theory to ferroelectric phase transformations, characterized by the non-centrosymmetric displacements of atoms in the ferroelectric phase, was accomplished by Ginzburg (Ginzburg 1945) and later adopted by Devonshire in a classical paper in 1949 to $BaTiO_3$ (Devonshire 1949; Devonshire 1951; Devonshire 1954). Ginzburg and Levanyuk incorporated the effect of spatially inhomogeneous thermal fluctuations of polarization near the phase transformation by expanding the Landau potential not only in terms of powers of the polarization but also of the gradients of the polarization (Levanyuk 1959; Ginzburg 1987). A correlation length was conceived that defines the region to which these polarization fluctuations would be confined and where the crystal would "feel" the ferroelectric phase transformation. These modifications constitute the basis of the so-called Landau-Ginzburg-Devonshire (LGD) thermodynamic theory of ferroelectric phase transformations. We will confine ourselves to the LGD formalism throughout this Chapter to explain the origin of the polarization offset in polarization-graded ferroelectrics as well as to understand the electrophysical response of these materials. The obvious advantage of LGD phenomenology is that a variety of physical properties can be explained without referring to all the degrees of freedom in complex materials systems such as ferroelectrics. For a complete analysis of the LGD approach and the physical foundations of ferroelectric phenomena, the reader is referred to the original article by Devonshire (Devonshire 1949; Devonshire 1951), excellent textbooks, notably by Strukov and Levanyuk (Strukov and Levanyuk 1998), Fatuzzo and Merz (Fatuzzo and Merz 1967), Jona and Shirane (Jona and Shirane 1962), and Lines and Glass (Lines and Glass 1977) and research/review papers by Cross and co-workers (Haun, Furman et al. 1987; Haun, Furman et al. 1989), and Damjanovic (Damjanovic 1998).

The major shortcoming of LGD thermodynamic theory is that it is not well-suited for dynamic processes such as polarization reversal (or switching) upon the application of an electric field in ferroelectric crystals (Drougard and Landauer 1959; Bratkovsky and Levanyuk 2000). Thermodynamic theory treats switching as an instability in one ferroelectric state with respect to the other under the application of an external electrical field. In reality, switching occurs via nucleation of reversed domains and their growth (Miller and Weinreich 1960). In Sec. 4.2 we will present a preliminary electrostatic analysis of a ferroelectric bilayer and analyze the stability of single-domain structure in the bilayer as a function of the strength of the depoling field. This field arises due to polarization mismatch at the interlayer interface and has a "symmetry-breaking" effect if the polarization mismatch is small or may completely suppress ferroelectricity in the layer with higher initial polarization. In the last Section of this Chapter, we will consider the electric field dependence of the polarization offset in polarization-graded ferroelectrics. A statistical mechanical approach will be presented based on skewed potential wells that favor one ferroelectric ground state over the other. We expect that the switching process is the result of domain wall motion in some respects quite similar to polarization reversal in homogeneous ferroelectrics but also quite different due to the presence of an inhomogeneous internal potential resulting from depolarization effects. We hope that this simple preliminary discussion will stimulate further research to understand the dynamics of the polarization offset.

2. LGD THEORY

The emergence of a spontaneous polarization and thus ferroelectricity in homogeneous ferroelectric materials is due to a spontaneous displacement of atoms relative to each other below a certain temperature creating permanent dipole moments. The density of such dipole moments is termed the polarization, which is a vectorial quantity. Most ferroelectric materials that are of practical interest have a perovskite structure with the chemical formula ABO_3. Some typical examples are $BaTiO_3$, $PbTiO_3$, $(Ba,Sr)TiO_3$ (BST), and lead-based solid solutions such as $Pb(Zr,Ti)O_3$ (PZT) and $(Pb,Sr)TiO_3$ (PST). Figure 2.1a shows the perovskite structure of $PbTiO_3$ above the transformation temperature where the TiO_6 octahedra are linked in a regular cubic array forming the high-symmetry $m3m$ prototype for many ferroelectric forms. The small 6-fold coordinated site in the center of the octahedron is filled by a small, highly-charged Ti^{+4} and the larger 12-fold coordinated "interstitial" site between octahedral carries a larger Pb^{+2}. The oxygen ions sit at the face-centers of the cubic cells. The spontaneous polarization in $PbTiO_3$ arises from the spontaneous non-centrosymmetric displacement of Ti^{+4} and O^{-2} ions relative to Pb^{+2} ions. Figure 2.1b shows an example of such a shift in the tetragonal phase, resulting a net polarization along the c-axis of the tetragonal unit cell. In general, the axis of polarization may be parallel to the unit cell edge, the face diagonal, or the body diagonal.

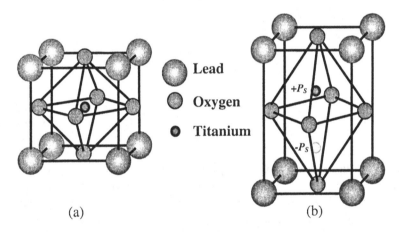

(a)   (b)

Figure 2.1. The perovskite crystal structure of $PbTiO_3$ (a) above, (b) below the ferroelectric phase transformation temperature. The tetragonality of the ferroelectric phase is highly exaggerated.

The polarization is accompanied by a self-strain. A spontaneous polarization along the cube edge results in an elongation along the direction of polarization and a corresponding contraction along directions perpendicular to it, reducing the symmetry to tetragonal ($4mm$). Similarly, polarization along the face diagonal reduces the symmetry to orthorhombic ($mm2$) and polarization along the body diagonal reduces it to rhombohedral ($3m$). $BaTiO_3$, which has been more studied than any other perovskite, undergoes these three phase transformations in succession as it is cooled from high temperatures, as shown in Figure 2.2. The high-symmetry, non-polar cubic phase transforms sequentially to low-symmetry ferroelectric phases with decreasing temperature.

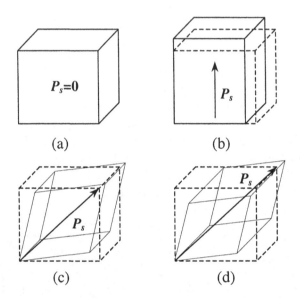

Figure 2.2. The crystal structure and the orientation of the spontaneous polarization direction of BaTiO$_3$ (a) above 130°C, cubic, (b) between 118°C and 5°C, tetragonal, (c) between 5°C and -90°C, (d) below -90°C, rhombohedral (Jona and Shirane 1962), .

Let us concentrate on the simplest ferroelectric transformation where upon cooling the paraelectric cubic *m3m* phase transforms to a ferroelectric tetragonal *4mm* phase. As can be seen from Figure 2.1b, the spontaneous polarization and the accompanying lattice distortion can be along the positive or the negative *z*-axis, or the (001) direction. For stress-free single-crystals, the (Helmholtz) free energy density can be described via an LGD potential that should be constructed in such a way that the existence of these two equivalent ferroelectric ground states are taken into account. This can be accomplished by the expanding the free energy (per unit volume) in terms of the polarization only with respect to even powers, such that:

$$F_L(P,T) = F_0 + \frac{1}{2}\alpha P^2 + \frac{1}{4}\beta P^4 + \frac{1}{6}\gamma P^6, \qquad (2.1)$$

where F_0 is the energy in the paraelectric state, P is the polarization (which is the order parameter of the phase transformation), and α, β, and γ are the expansion (or the dielectric stiffness) coefficients. α is a temperature dependent coefficient and its dependency is given by the Curie-Weiss law:

$$\alpha = \frac{T - T_C}{\varepsilon_0 C}, \qquad (2.2)$$

where T_C and C are the Curie-Weiss temperature and constant, respectively, and ε_0 is the permittivity of free space. The other two expansion coefficients are usually taken to be independent of temperature.

The Landau potential should exhibit only one minima at $P=0$ above T_C corresponding to the centrosymmetric cubic non-polar phase. Below T_C, the free energy curve should have two minima corresponding to two identical ferroelectric ground states but with opposite orientation of the polarization direction along the "easy" (001) orientation (P_S and $-P_S$). The schematic free energy potentials at various critical temperatures are illustrated in Figure 2.3 for a 2nd-order and Figure 2.4 for a 1st-order ferroelectric phase transformation. It can be shown that if $\beta<0$, the phase transformation from the paraelectric state is of 1st-order (i.e., discontinuity in P_S, and lattice parameters at T_C and thermal hysteresis in the same parameters around T_C) and it is of 2nd-order if $\beta>0$ (i.e., gradual variation in P_S, and lattice parameters below T_C with no thermal hysteresis).

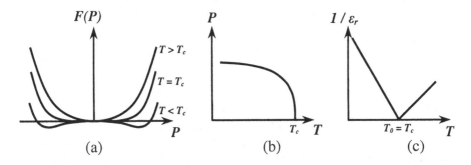

Figure 2.3. Second-order transition: (a) Free Energy versus polarization, (b) spontaneous polarization versus temperature, (c) reciprocal susceptibility versus temperature (Jona and Shirane 1962).

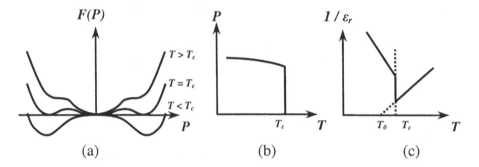

Figure 2.4. First-order transition: (a) Free Energy versus polarization, (b) spontaneous polarization versus temperature, (c) reciprocal susceptibility versus temperature .

The spontaneous polarization P_S in the tetragonal phase can be obtained from the condition for thermodynamic equilibrium $\partial F_L / \partial P = 0$ such that:

$$P_S^2(T) = \frac{-\beta + (\beta^2 - 4\alpha\gamma)^{1/2}}{2\gamma}. \tag{2.3}$$

The spontaneous polarization is due to the non-centrosymmetric displacements the ions in the crystal and results in a tetragonal distortion. The structural component of the phase transformation can be described by the self-strains. It should be noted that the self-strain is not actually a strain *per se* but rather a geometric description of the lattice distortion. The components of the self-strain tensor are defined as:

$$x_1^0 = x_2^0 = \frac{a - a_0}{a_0} = Q_{12}P_S^2 , \quad x_3^0 = \frac{c - a_0}{a_0} = Q_{11}P_S^2 , \tag{2.4}$$

where a and c are the lattice parameters in the tetragonal ferroelectric state and a_0 is the lattice parameter of the cubic phase. Q_{ij} is a 4th-rank tensor (in the contracted notation that we will use throughout this book) that couples the polarization with the self-strain, called the electrostrictive coefficients. We present in Figures 2.5 and 2.6 the variation of the spontaneous polarization and the lattice parameters of PbTiO$_3$ as a function of temperature, respectively, based on the experimentally determined expansion parameters and physical constants as reported in the literature (Haun, Furman et al. 1987).

Obviously, for the tetragonal state, there are 3 orientational variants defined by the self-strain tensors:

$$\varepsilon_1^0 = \begin{pmatrix} x_3^0 & 0 & 0 \\ 0 & x_1^0 & 0 \\ 0 & 0 & x_1^0 \end{pmatrix}, \varepsilon_2^0 = \begin{pmatrix} x_1^0 & 0 & 0 \\ 0 & x_3^0 & 0 \\ 0 & 0 & x_1^0 \end{pmatrix}, \varepsilon_3^0 = \begin{pmatrix} x_1^0 & 0 & 0 \\ 0 & x_1^0 & 0 \\ 0 & 0 & x_3^0 \end{pmatrix}. \quad (2.5)$$

Noting that the direction of the spontaneous polarization can be along two opposite directions along the easy axis, we conclude that there are 6 electrical variants and 3 structural variants, or domains (see Figure 2.7). In other words, there are two ferroelectric ground states for each one of the orientational domains. Of course, the number of electrical and structural domains increases as the symmetry of the ferroelectric phase decreases. For example, there are 6 structural and 12 electrical domains in orthorhombic $BaTiO_3$.

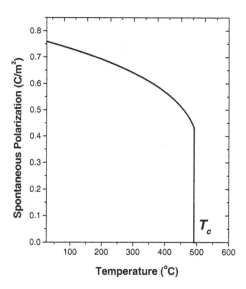

Figure 2.5. The variation in the spontaneous polarization of $PbTiO_3$ as a function of temperature calculated from phenomenological parameters.

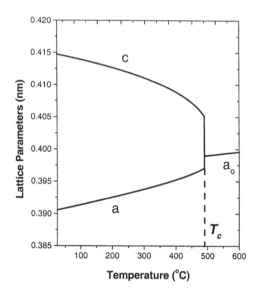

Figure 2.6. The variation in the lattice parameters of PbTiO$_3$ as a function of temperature calculated from phenomenological parameters.

The effect of applied stresses and electric fields can be incorporated into the Landau potential. The energy of elastic stresses in a pseudo-cubic crystal is given by:

$$F_\sigma = \frac{1}{2} S_{11}(\sigma_1^2 + \sigma_2^2 + \sigma_3^2) + S_{12}(\sigma_1\sigma_2 + \sigma_1\sigma_3 + \sigma_2\sigma_3)$$
$$+ \frac{1}{2} S_{44}(\sigma_4^2 + \sigma_5^2 + \sigma_6^2)$$

(2.6)

where S_{ij} are the elastic compliances at constant polarization and σ_i components of the applied elastic stress tensor. It should be kept in mind that the applied stress is coupled with the polarization due to the electrostrictive effect and the energy of this coupling can be expressed by:

$$F_C = x_1^0(\sigma_1 + \sigma_2) + x_3^0\sigma_3 = Q_{12}P^2(\sigma_1 + \sigma_2) + Q_{11}\sigma_3 P^2.$$ (2.7)

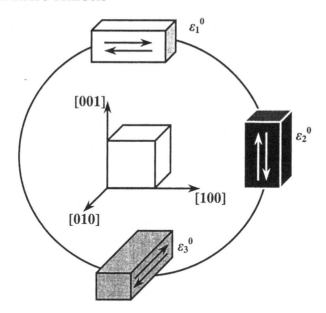

Figure 2.7. Three orientational variants (or elastic domains) of the tetragonal ferroelectric phase. The initial cubic paraelectric phase is shown at the center as a reference.

The energy contribution of an applied electric field is given by:

$$F_E = E \cdot P \qquad (2.8)$$

where E is the electric field along the easy axis.

Taking into account these contributions, the total energy (per unit volume) is described by the Gibbs free energy:

$$G(P,T,\sigma_i,E) = F_L - F_\sigma - F_C - F_E . \qquad (2.9)$$

We note that the proper description of the (total) free energy can only be done using the (elastic) Gibbs function as the Helmholtz free energy describes the internal energy of the system (in the absence of external fields).

The electrical and electromechanical properties of a ferroelectric can then be obtained by taking the appropriate derivatives of the above free energy functional. For example, the dielectric susceptibility along the easy axis under zero stress is given by $(\partial^2 G/\partial P^2)^{-1}$ such that:

$$\chi_3 = (\alpha + 3\beta P^2 + 5\gamma P^4)^{-1}, \qquad (2.10)$$

and the relative dielectric constant along the easy axis is defined as:

$$\varepsilon_r \varepsilon_0 = 1 + \chi_3 \text{ or } \varepsilon_r \cong \chi_3 / \varepsilon_0, \tag{2.11}$$

since $\chi_3 >> 1$.

The (converse) piezoelectric coefficient that describes the strain due to an applied electric field along the (001) direction is defined through the equation, $d_{33} = \chi_3 (\partial^2 G/\partial P \partial \sigma_3)$, such that:

$$d_{33} = \varepsilon_0 \chi_3 Q_{11} P. \tag{2.12}$$

The pyroelectric response that describes the variation in the spontaneous and induced polarization (or the dielectric displacement) with temperature is given by:

$$p = \frac{\partial P_S}{\partial T} + E \frac{\partial \varepsilon_3}{\partial T}, \tag{2.13}$$

where the last term describes the temperature dependence of polarization induced by an applied (external) electric field.

We note that the Landau potential can be somewhat more complex for the low temperature phases of $BaTiO_3$ where the ionic displacement (or correspondingly the polarization) can be along the (110) directions in the orthorhombic phase or along the (111) directions in the rhombohedral phase. For these polymorphs, the expansion of the free energy has to be carried out for multiple order parameters; two for the orthorhombic and three for the rhombohedral phase (Devonshire 1949; Devonshire 1951). Additional terms may arise if there are structural variations due to the tilting or rotation of the oxygen octahedra as it is the case for the "incipient" ferroelectric $SrTiO_3$ (Slonczewski and Thomas 1970; Pertsev, Tagantsev et al. 2000). In such cases, two coupled sets of order parameters, one set describing displacements and the other set describing rotations, may have to be employed (Salje 1990).

The Landau formalism is based on the assumption that the polarization of the system is homogenous and there are no (local) variations in the order parameter. This is valid at relatively low temperatures where fluctuations of the order parameter are small. However, thermal vibrations close to the phase transformation temperature T_C may result in strong inhomogeneities in the polarization, especially in the short range. Following Strukov and Levanyuk (Strukov and Levanyuk 1998) the mean square deviation $<P^2>$ of the polarization due to thermal fluctuations is given by:

$$<P^2> = \int_{-\infty}^{\infty} P^2 \exp\left(-\frac{\omega F_L(P,T)}{k_B T}\right) dP \left[\int_{-\infty}^{\infty} \exp\left(-\frac{\omega F_L(P,T)}{k_B T}\right) dP\right]^{-1}, \quad (2.14)$$

where ω is the volume of the crystal and k_B is the Boltzmann constant. In the symmetric paraelectric phase ($P=0$), $F_L=F_0+1/2\alpha P^2$, and with $F_0=0$, the above integral can be evaluated to be:

$$<P^2> = \frac{\varepsilon_0 C}{\omega} \frac{k_B T}{(T-T_C)}, \quad (2.15)$$

It can be immediately seen that $<P^2> \to \infty$ as $\Delta T=(T-T_C) \to 0$, indicating the presence of spatially inhomogeneous fluctuations in the polarization near the phase transformation temperature, occurring over a small characteristic length. The additional energy of the crystal due to these thermal fluctuations can then be represented as the integration of the volume energies of small regions with uniform polarization (given by F_L). As shown by Ginzburg (Ginzburg 1987), the energy interaction between these small regions with different polarizations can be approximated by:

$$F_G = \frac{1}{2} D \left(\frac{dP}{dz}\right)^2 \approx \frac{1}{2} \delta^2 |\alpha| \left(\frac{dP}{dz}\right)^2 \quad (2.16)$$

for fluctuations in one dimension, where δ is a characteristic length, of the order of the linear dimensions of the region size.

The revised Landau potential that incorporates these (one-dimensional) polarization inhomogeneities is thus given by:

$$F_L(P,T) = F_0 + \frac{1}{2}\alpha(T)P^2 + \frac{1}{4}\beta P^4 + \frac{1}{6}\gamma P^6 + \frac{1}{2}D(T)\left(\frac{dP}{dz}\right)^2. \quad (2.17)$$

The above relation can also be obtained on the basis of considerations associated with the possibility of expanding the thermodynamic potential into a series in powers not only of P but of the derivatives of P as well. The parameter D is always (taken to be) positive, thus the gradient term in the above relation acts as a restoring force that serves to damp out the spatial variations in P.

3.   FERROELECTRIC HYSTERESIS

One of the most important features of ferroelectrics is the reversal of the polarization that involves the switching from one ferroelectric ground state to the other in the presence of an applied electric field. As pointed out earlier, the polarization (or the relative displacements of ions) in a tetragonal ferroelectric can be along the positive or negative z-axis (see Figure 2.1). These two identical ferroelectric ground states are defined by the condition of equilibrium, $\partial G/\partial P=0$ at $P=P_S$ and $P=-P_S$ (Figure 2.3). The formation of these electrical variants is obviously equally probable and usually results in the formation of regions (or domains) with opposite direction of polarization, yielding zero net polarization in a "virgin" ferroelectric.

If a uniform electric field is applied along the positive z-axis, the volume fraction of the domains that are favorably oriented with respect to the field increases via domain wall motion giving rise to a measurable net polarization. The polarization behavior is non-linear as shown in Figure 1.4. Once all the domains have "switched," further increase in the applied field results in a proportional increase in the polarization until a saturation value is reached. Upon the removal of the applied field, the ferroelectric does not revert to its initial state although some domains do switch back. The value of the polarization at zero field after poling is called the remanent polarization (P_R). The subsequent application of a uniform field along the negative z-axis increases the volume fraction of domains with $P=-P_S$. At a critical field $-E_C$, called the coercive field, the volume fractions of the domains become equal and the net polarization is zero. Again, an increase in the negative field eventually yields a single-domain state with the polarization pointing in the negative z-direction. Upon removing the negative field, the polarization of the ferroelectric reverts to $-P_R$. The above described process results in a hysteresis corresponding to an energy loss. This energy is essentially spent to nucleate and grow domains with the opposite polarization.

Let us theoretically analyze the effect of a uniform electric field applied along the easy polarization axis of an unclamped tetragonal ferroelectric single-crystal via the LGD theory. Far away from the phase transformation temperature where thermal fluctuations in the polarization are insignificant, the energy of the system per unit volume in the absence of external stress fields is given by:

$$G(P,T,E) = F_0 + \frac{1}{2}\alpha P^2 + \frac{1}{4}\beta P^4 + \frac{1}{6}\gamma P^6 - E \cdot P \qquad (2.18)$$

Minimization with respect to the polarization yields:

$$\frac{\partial G}{\partial P} = 0, \quad E = \alpha P + \beta P^3 + \gamma P^5 \qquad (2.19)$$

The above relation defines the equilibrium polarization of a ferroelectric in the presence of an applied electric field. In Figure 2.8, we plot the theoretical polarization as a function of the applied field in an unconstrained PbTiO$_3$ single-crystal. In the region A-B, it is clear that the second derivative of the Gibbs function is negative, indicating instability. This means that at A, defined by a critical field $-E_C$, the ferroelectric ground state along the positive z-axis becomes unstable. Similarly, at point B the ferroelectric with polarization along the negative z-axis becomes unstable at E_C. Thus, we obtain the hysteresis behavior observed experimentally with one significant difference. The theoretically predicted value of the coercive field may be two- to three-orders of magnitude larger than the experimentally observed values. This is because of the fact that switching in the LGD treatment is due to thermodynamic instability of the polarization with respect to an applied field in the reverse direction rather than the nucleation and growth of electrical domains.

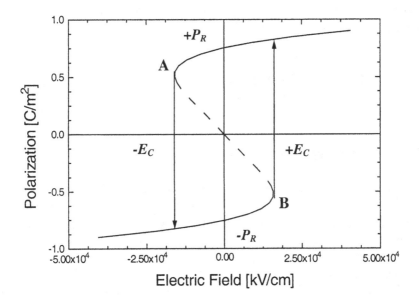

Figure 2.8. Theoretical polarization versus electrical field curve for PbTiO$_3$, calculated from phenomenological parameters.

Although there have been many efforts to theoretically describe the polarization reversal based on domain wall motion and nucleation and growth of electrical domains, electrostatic considerations, Landau models, or micro-electrical models among others (Ishibashi 1992; Shur and Rumyantsev 1994; Shur, Rumyantsev et al. 1994; Tagantsev, Pawlaczyk et al. 1994; Shur, Rumyantsev et al. 1995; Tagantsev, Pawlaczyk et al. 1995; Chai, Brews et al. 1997; Chen, Fang et al. 1997; Shur, Ponomarev et al. 1997; Chen and Lynch 1998; Hwang, Huber et al. 1998; Ricinschi, Harnagea et al. 1998; Bartic, Wouters et al. 2001; Robert, Damjanovic et al. 2001), the modeling of the switching and hysteresis of ferroelectrics is still an issue that remains unsolved. The major difficulty in establishing a theoretical description is related to the ambiguity of the shape of the nuclei of reversed domains and their sidewise motion. Compared to other nucleation and growth phenomena in the solid state [such as solidification or precipitation of second-phase particles in an alloy (Christian 1975; Kostorz (ed.) 2001)], in ferroelectrics the switching process occurs much faster, in nanoseconds or less. This obviously complicates experimental observations, which would provide a starting point for the theoretical analysis. Most notably, models based on the classical Kolmogorov-Avrami approach have been adapted to ferroelectrics (Ishibashi and Takagi 1971; Ishibashi and Orihara 1986; Ishibashi 1992; Ishibashi and Orihara 1992; Orihara and Ishibashi 1992; Orihara, Hashimoto et al. 1994) but much remains to be understood.

4.   LGD THEORY FOR GRADED FERROELECTRICS

4.1   Depoling due to Polarization Fluctuations

Variations in the polarization within a volume result in internal electric fields that may be strong enough to suppress the polarization in the given volume (Bratkovsky and Levanyuk 2002; Glinchuk, Eliseev et al. 2002; Glinchuk, Eliseev et al. 2003). This depoling field can be calculated using basic electrostatic considerations. Confining ourselves to spatial polarization variations only along a single direction defined by a z-axis in a uniaxial ferroelectric such that $\mathbf{P}=[0,0,P(z)]$, the dielectric displacement $\mathbf{D}=[0,0,D(z)]$ due to a depoling field E_D is given by:

$$D(z) = \varepsilon_0 E_D(z) + P(z). \qquad (2.20)$$

The polarization has two components: the spontaneous polarization of the ferroelectric and the electric field induced polarization. Taking into account the possibility of having free charges in the system, we write the one-dimensional Poisson's relation as:

$$\frac{\partial D(z)}{\partial z} = \rho, \qquad (2.21)$$

where ρ is the free charge density, and thus:

$$\frac{\partial E_D(z)}{\partial z} = \frac{1}{\varepsilon_0}\left[\rho - \frac{\partial P(z)}{\partial z}\right]. \qquad (2.22)$$

The polarization gradient is the source of the depoling field and can be compensated by the flow of free charges within the material (due to the small but finite conductivity) or by free charges from the surrounding medium. We realize that the free charge density may be a complex function of position (Fridkin 1980). However, for the current analysis what is of most interest is the magnitude of the total number of free charges that may be present in the graded ferroelectrics.

Let us consider a ferroelectric material that is an insulator with no free charges to compensate the depoling field due to variations in the polarization in one direction z. In this case, we have $\partial D/\partial z = 0$ and thus:

$$\frac{dE_D}{dz} = -\frac{1}{\varepsilon_0}\cdot\frac{dP}{dz}. \qquad (2.23)$$

For a parallel-plate ferroelectric with in-plane dimensions much larger than the thickness L along which there are polarization fluctuations, the solution for the above relation employing proper boundary conditions is given by (Kretschmer and Binder 1979):

$$E_D(z) = -\frac{1}{\varepsilon_0}\left[P(z) - \frac{1}{L}\int_0^L P(z)dz\right] = -\frac{1}{\varepsilon_0}[P(z) - <P>], \qquad (2.24)$$

where L is the length along which there is a polarization inhomogeneity and
<P> is the average polarization. The energy density of the depolarization is
then given by:

$$F_D = -\frac{1}{2}\xi E_D(z) \cdot P(z) = -\frac{\xi}{2\varepsilon_0}[P(z) - <P>] \cdot P(z), \qquad (2.25)$$

where we have introduced a parameter $0 \leq \xi \leq 1$ which is a measure of the
strength of the depoling field. $\xi=1$ corresponds to the case of a perfect
insulating ferroelectric crystal in vacuum and $\xi=0$ is the case of a
semiconductor ferroelectric that contains enough free charges with high
mobility to compensate for the depoling field. Physically, ξ is the ratio of the
free charge density to bound charge density given by:

$$\xi - 1 = \frac{\rho L}{P - <P>}, \qquad (2.26)$$

where L is the length of the ferroelectric bar in the direction of the
polarization gradient. We will discuss the physical meaning of this coefficient
and its implications in detail in the next sections.

4.2   Electrical Domains and Interlayer Coupling in Layered Ferroelectrics

Before we continue with our thermodynamic analysis, it is necessary to
clarify an important aspect of our analysis. The contribution of the depoling
field to the total energy of an inhomogeneous ferroelectric can be analyzed for
the two extreme cases; $\xi=1$ corresponding to a perfect insulating ferroelectric
bar in vacuum, and $\xi \approx 0$ corresponding to almost complete charge
compensation. For a polarization-graded ferroelectric with good electrical
insulating properties, a simple estimation employing experimentally
determined expansion parameters for well-known materials systems shows
that the energy of the depoling field is approximately two orders of magnitude
larger than the Landau expansion terms, resulting in an instability that should
completely suppress ferroelectricity within each and every "layer." A stable
polarization within each "layer" can only be achieved if $\xi=0.01-0.001$. This
means that there should be an extremely large number of free charges
corresponding to $10^{17}-10^{20}$ cm^{-3}, which obviously cannot be the case for most
ferroelectrics as they are good electrical insulators that find numerous uses in

technological applications as capacitor elements. Even in the case of ferroelectric thin films, which contain a host of charged point defects, this much conductivity cannot easily be realized.

The above conclusion is, of course, at odds with experimental results, which will be summarized, in the next chapters. It is shown that polarization-graded ferroelectrics not only display well-defined hysteresis in their polarization vs. applied electric field behavior (indicating stability and reversibility of two ferroelectric ground states), but also an offset in the polarization as the applied field increases.

The answer to this apparent paradox can be explained for two limiting cases. For steep polarization gradients; i.e., a relatively large ratio of variation of polarization to the total length of the ferroelectric bar along which there is the polarization gradient, the depoling field can be minimized by the formation of electrical domains. For relatively smooth polarization gradients, a single-domain state in each and every layer can still be maintained without depoling. We will provide a qualitative discussion for both these cases and we develop a simple analysis for the condition of single-domain stability in a ferroelectric bilayer.

Bratkovsky and Levanyuk studied ferroelectric bilayers using a thermodynamic model which took into account electrostatic interactions (Bratkovsky and Levanyuk 2002). They showed that slight inhomogeneities in the polarization of insulating multilayer ferroelectric media should result in the formation of electrical domains, which minimize the depoling field. As a continuously graded ferroelectric is the limiting case of this analysis, we may extend their findings for steep polarization gradients and conclude that the depoling field associated with the overall polarization gradient is significantly reduced and may even be completely eliminated by the formation of electrical domains.

It is thus clear that although the "global" polarization gradient is marginalized or completely eliminated, a local gradient in the polarization still exists. Electrical domains in homogeneous ferroelectrics consist of regions that are polarized with the direction of the polarization reversed as one moves from one domain to the other. In a ferroelectric plate without top and bottom electrodes, the formation of electrical (or 180°) domains results in a checkerboard pattern as shown in Figure 2.9. This configuration decreases the depoling field and the net polarization of the ferroelectric is obviously zero in the absence of an applied field (Kittel 1946; Forsbergh 1949; Merz 1954; Kooy and Enz 1960). The poling of such a structure is accomplished by sidewise domain wall motion in the presence of an applied electric field until domains with anti-parallel orientation are consumed. Reversal of the polarization occurs through nucleation and growth of anti-parallel domains

(Miller and Weinreich 1960). In polarization-graded ferroelectrics (or even in a simple ferroelectric bilayer), a similar behavior is expected although the domain pattern that minimizes the depoling field will necessarily be much more complicated than for the single-layer ferroelectric shown in Figure 2.9. We note that not much is known as to the kind of a domain configuration formed in polarization-graded ferroelectrics at this time. It is clear, however, that the overall depoling field would be reduced as to preserve ferroelectricity. Although in the absence of an applied field, globally *dP/dz*=0, locally however on a length scale less than the domain period, *dP/dz* cannot be equal to zero. This local polarization variation is the source of the built-in potential.

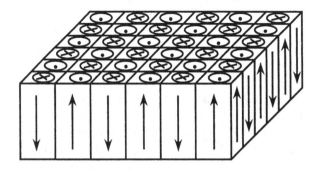

Figure 2.9. Electrical (or 180°) domains in a homogenous ferroelectric that minimize the depoling energy.

Can a single-domain state in a multilayer ferroelectric be stabilized and if so, what are the conditions? This is a question that we will try to answer quantitatively. Consider two freestanding ferroelectric layers with equal lateral dimensions as shown in Figure 2.10a. Both ferroelectric layers have a perovskite structure, with the spontaneous polarization along the *z*-direction in the ferroelectric state. Assuming that the paraelectric-ferroelectric phase transformation in each layer is second-order, we can express the energy per unit volume of the layers in an uncoupled, unconstrained condition only up to the fourth power of polarization as:

$$F_1 = F_{0,1} + \frac{1}{2}aP_1^2 + \frac{1}{4}bP_1^4$$

$$F_2 = F_{0,2} + \frac{1}{2}cP_2^2 + \frac{1}{4}dP_2^4$$

(2.27)

where $F_{0,i}$ is the energy of layer i in the paraelectric state, and a, b, c, and d are Landau coefficients. a and d are temperature dependent and their temperature dependency is given by the Curie-Weiss law, i.e., $a=T-T_{C,1}/\varepsilon_0 C_1$ and $c=T-T_{C,2}/\varepsilon_0 C_2$ where ε_0 is the permittivity of free space, $T_{C,i}$ and C_i are the Curie-Weiss temperature and constant of layer i. The other coefficients for both materials are assumed to be temperature independent. The equilibrium spontaneous polarization in each layer is given by the equations of state such that:

$$\frac{\partial F_1}{\partial P_1} = 0, \; P_{1,0} = \pm\sqrt{-\frac{a}{b}} \qquad (2.28)$$

$$\frac{\partial F_2}{\partial P_2} = 0, \; P_{2,0} = \pm\sqrt{-\frac{c}{d}} \qquad (2.29)$$

Suppose that ferroelectric 1 with thickness h_1 is joined together with ferroelectric 2 with thickness h_2 as shown in Figure 2.10b. Their relative volume fraction α can be expressed through the thickness of each layer:

$$\alpha = \frac{h_2}{h_1 + h_2} = \frac{h_2}{h} . \qquad (2.30)$$

Before layer 1 and layer 2 are brought together, their polarizations are equal to their bulk values $P_{1,0}$ and $P_{2,0}$. We assume that $P_{1,0} > P_{2,0}$ (see Figure 2.10a). Obviously, when these layers are coupled the actual polarization in each layer is expected to be different from its "decoupled" bulk value due to the electrical interaction between the layers. The internal electric fields $E_{D,1}$ and $E_{D,2}$ in layer 1 and layer 2 due to the polarization mismatch essentially establish a new polarization state, i.e., P_1 and P_2 in layers 1 and 2, respectively (Figure 2.10b). The internal electric fields are related to the difference of polarization in each layer and are given by:

$$E_{D,1} = -\frac{1}{\varepsilon_0}\left(P_1 - <P>\right) = \frac{\alpha}{\varepsilon_0}(P_2 - P_1) \qquad (2.31)$$

$$E_{D,2} = -\frac{1}{\varepsilon_0}\left(P_2 - <P>\right) = \frac{1-\alpha}{\varepsilon_0}(P_1 - P_2) \qquad (2.32)$$

where

$$< P >= (1 - \alpha)P_1 + \alpha P_2 \qquad (2.33)$$

is the average polarization.

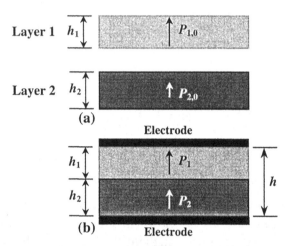

Figure 2.10. Schematics of a ferroelectric bilayer: (a) Uncoupled, (b) Coupled.

It is clear that the role of the internal depoling field is quite different in layer 1 and 2. The electric field in layer 1 $E_{D,1}$ attempts to decrease the polarization since it lies anti-parallel with the polarization vector ($E_{D,1}<0$). It is anticipated, therefore, that if $E_{D,1}$ is higher than the theoretical coercive field of layer 1 ($E_{C,1}$), it may completely switch the polarization in layer 1, leading to the instability of a single-domain state. On the other hand, the field in layer 2 serves to enhance the polarization in this layer ($E_{D,2}>0$) and thus a single-domain ferroelectric state in layer 2 will always be stable.

We will now show how the internal electric field can be estimated from the initial bulk polarization difference between the layers and analyze the condition for the stability of a single-domain state in both layers. The basic thermodynamic potentials that relate the electric field and the polarization, assuming P^4 approximation, can be written as:

$$\frac{\partial F_1}{\partial P_1} = aP_1 + bP_1^3 = \frac{\alpha}{\varepsilon_0}(P_2 - P_1) = E_{D,1} \qquad (2.34)$$

$$\frac{\partial F_2}{\partial P_2} = cP_2 + dP_2^3 = \frac{1-\alpha}{\varepsilon_0}(P_1 - P_2) = E_{D,2} \qquad (2.35)$$

The bulk spontaneous polarization for layer 1 and 2 ($P_{1,0}$ and $P_{2,0}$, respectively) satisfy the equation of state in the absence of any internal or external field, i.e.:

$$aP_{1,0} + bP_{1,0}^3 = 0 \qquad (2.36)$$

$$cP_{2,0} + dP_{2,0}^3 = 0 \qquad (2.37)$$

Combination of Equation 2.34 and 2.36 yields:

$$a(P_1 - P_{1,0}) + b(P_1^3 - P_{1,0}^3) = \frac{\alpha}{\varepsilon_0}(P_2 - P_1) \qquad (2.38)$$

or:

$$(P_1 - P_{1,0})\left[a + b(P_1^2 + P_1 P_{1,0} + P_{1,0}^2)\right] = \frac{\alpha}{\varepsilon_0}(P_2 - P_1), \qquad (2.39)$$

which can be approximated as

$$(P_1 - P_{1,0})(a + 3bP_{1,0}^2) = \frac{\alpha}{\varepsilon_0}(P_2 - P_1). \qquad (2.40)$$

With $P_{1,0}^2 = -a/b$, we get:

$$-2a(P_1 - P_{1,0}) = \frac{\alpha}{\varepsilon_0}(P_2 - P_1). \qquad (2.41)$$

Similarly,

$$-2c(P_2 - P_{2,0}) = \frac{1-\alpha}{\varepsilon_0}(P_1 - P_2) \qquad (2.42)$$

Combination of Equations 2.41 and 2.42 leads to:

$$P_1 - P_2 = \frac{P_{1,0} - P_{2,0}}{1 - \frac{1}{2}\left[\frac{\alpha}{a\varepsilon_0} + \frac{1-\alpha}{c\varepsilon_0}\right]} \tag{2.43}$$

The Landau coefficients a and c can be related to the relative dielectric constant $\varepsilon_{r,1}$ and $\varepsilon_{r,2}$ through

$$\varepsilon_{r,1} = \frac{\varepsilon_1}{\varepsilon_0} = \frac{1}{\varepsilon_0}\left(\frac{\partial^2 F_1}{\partial P_1^2}\right)^{-1} = \frac{1}{\varepsilon_0(a + 3bP_{1,0}^2)} = -\frac{1}{2a\varepsilon_0} \tag{2.44}$$

$$\varepsilon_{r,2} = \frac{\varepsilon_2}{\varepsilon_0} = \frac{1}{\varepsilon_0}\left(\frac{\partial^2 F_2}{\partial P_2^2}\right)^{-1} = \frac{1}{\varepsilon_0(c + 3dP_{2,0}^2)} = -\frac{1}{2c\varepsilon_0} \tag{2.45}$$

Therefore, the polarization jump in bilayer P_1-P_2 can be obtained by knowing solely the polarization difference of the layers in bulk $P_{1,0}$-$P_{2,0}$, and their bulk dielectric constants as follows:

$$P_1 - P_2 = \frac{P_{1,0} - P_{2,0}}{1 + [\alpha\varepsilon_{r,1} + (1-\alpha)\varepsilon_{r,2}]} \tag{2.46}$$

The bulk spontaneous polarization for $BaTiO_3$ and $PbTiO_3$ are 27 and 76 $\mu C/cm^2$, respectively. Their relative dielectric constants are 200 and 115. The resulting polarization difference in a bilayer system with equal volume fraction of $BaTiO_3$ and $PbTiO_3$ is estimated to be 0.3 $\mu C/cm^2$, which is more than two orders of magnitude less than the polarization difference between bulk $BaTiO_3$ and $PbTiO_3$. This small difference, however, may still result in a large internal field ($\sim 1.7\times10^3$ kV/cm) in the $PbTiO_3$ layer as determined by Equation 2.31, which is larger than the theoretical coercive field of bulk PT ($\sim 1.6\times10^3$ kV/cm). This suggests that a single domain state in PT layer should not be expected because of the large depolarization field. If we carry out the same analysis for an equi-fraction $BaTiO_3$-$Ba_{0.90}Pb_{0.10}TiO_3$ bilayer, we get an internal field of $\sim$141 kV/cm which is slightly less than the theoretical coercive field of $BaTiO_3$ (150 kV/cm). The internal field for a $BaTiO_3$-$Ba_{0.99}Pb_{0.01}TiO_3$ is even less, around 15 kV/cm. In both cases, the small polarization jump results in a depolarizing electric field that is less than the theoretical coercive strength of the material with the higher "uncoupled" polarization. Thus, it is expected that both layers should have a single-domain state. For more accurate results, Equations 2.34 and 2.35 should be solved numerically, taking into account the sixth order polarization terms.

Introducing ξ as defined in the previous section, the effect of free charges can be incorporated into the above analysis. The depoling field in layer 1 and layer 2 are now given by:

$$E_{D,1} = \frac{\alpha\xi}{\varepsilon_0}(P_2 - P_1), \tag{2.47}$$

$$E_{D,2} = \frac{(1-\alpha)\xi}{\varepsilon_0}(P_1 - P_2). \tag{2.48}$$

Using a similar approach, we get:

$$P_1 - P_2 = \frac{P_{1,0} - P_{2,0}}{1 + \xi[\alpha\varepsilon_{r,1} + (1-\alpha)\varepsilon_{r,2}]}. \tag{2.49}$$

Using this relation, we can calculate a critical ξ_C corresponding to the onset of the instability of single domain state given by:

$$E_{D,1} = \frac{\alpha\xi^*}{\varepsilon_0}(P_2 - P_1) = E_{C,1}. \tag{2.50}$$

Thus:

$$\xi^* = \frac{1}{\alpha\dfrac{P_{1,0} - P_{2,0}}{E_{C,1}\varepsilon_0} - [\alpha\varepsilon_{r,1} + (1-\alpha)\varepsilon_{r,2}]}. \tag{2.51}$$

ξ^* can be estimated to be ~0.006 for an equal fraction BaTiO$_3$-PbTiO$_3$ bilayer. If ξ is larger than ξ^*, there is a ferroelectric instability in the layer having the larger polarization; and conversely for ξ smaller than ξ^*, a single-domain state may be realized in both layers.

Alternatively, for a given ξ, the difference between the bulk polarization of layer 1 and 2 should be:

$$(P_{1,0} - P_{2,0})^* = \frac{E_{C,1}\varepsilon_0[\alpha\varepsilon_{r,1} + (1-\alpha)\varepsilon_{r,2}]}{\alpha\xi} \tag{2.52}$$

for a single-domain state to be stable in both layers.

This simple analysis shows that a single-domain state can be maintained in a ferroelectric bilayer if the initial uncoupled polarization difference between the layers is relatively small with the assumption that both layers are perfect insulators. The polarization difference can be larger if there are free charges in the ferroelectric layers. This shows that a polarization gradient may not result in suppression of ferroelectricity or in the formation of electrical domains if the polarization gradient is relatively smooth.

4.3　　Internal Stresses due to a Polarization Gradient

Similar to the depoling field, there is a built-in, position dependent stress field within a ferroelectric with polarization variations. This is due the electrostrictive coupling between the polarization and the self-strain. Consider a compositionally or temperature graded unconstrained single-crystal ferroelectric bar along the z-direction (see Figure 1.8). This bar may be thought of being composed of individual "layers" with a uniform polarization (and thus self-strain) along the z-direction as shown in Figure 2.11. There exists a biaxial stress state with equal orthogonal components in the xy-plane of each layer and the corresponding mechanical boundary conditions are given by $\sigma_1 = \sigma_2$, and $\sigma_3 = \sigma_4 = \sigma_5 = \sigma_6 = 0$, where σ_i are the components of the internal stress tensor. Thus, the total internal elastic energy due to the self-strain gradient for each "layer" can then be expressed as:

$$F_{el} = \frac{1}{2}(\sigma_1 x_1 + \sigma_2 x_2),\tag{2.53}$$

where $x_1 = x_2$, and $\sigma_1 = \sigma_2 = \sigma$ are the internal stresses and strains, respectively, in the xy-plane. We note that although the out-of-plane strain $x_3 \neq 0$, it will not have an effect on the strain energy since the material is unconstrained in the z-direction and hence $\sigma_3 = 0$. The in-plane strain is determined by the condition that both the average internal stress and the average momentum of the internal stress should be zero, i.e.:

$$\int_0^L \sigma(z)dz = 0, \qquad \int_0^L z\sigma(z)dz = 0.\tag{2.54}$$

From these relations follows that the strain $x_1 = x_2$ in each individual layer is given by (Roytburd and Slutsker 2002):

$$x_1(z) = x_2(z) = x^0(z) + \left(z - \frac{L}{2}\right)\frac{\partial^2 w}{\partial x^2} = x^0(z) + \left(z - \frac{L}{2}\right)\kappa, \qquad (2.55)$$

where w is the out-of-plane displacement. The first term represents the variation of the misfit in the xy-plane between a particular layer and the average self-strain $Q_{12}<P>^2$:

$$x^0(z) = Q_{12}[P^2(z) - <P>^2]. \qquad (2.56)$$

The layer that has a polarization equivalent to the average polarization is distinguished by the dark gray color in Figure 2.11.

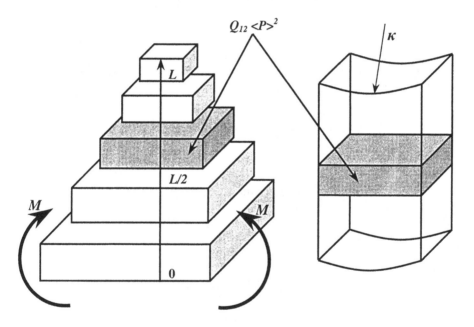

Figure 2.11. Schematic diagrams showing the one-dimesional variation in the misfit between a particular "layer" and the layer with average self strain given by Equation 2.56. The self-strain results in a bending moment M

Equation 2.56 has a similar form to the relation for the depoling field given by Equation 2.24. Both relations describe the additional internal energy due to the deviation of the polarization from the average. It is clear that that these contributions would not exist for a homogeneous polarization distribution where $P(z)=<P>=P_S$.

Unlike the case for the depoling field, there is also an additional elastic energy associated with the bending of the ferroelectric due to an uncompensated moment (Roytburd and Slutsker 2002) given by the second term in Equation 2.55 with a radius of curvature:

$$\kappa = \frac{24}{L^3} \int_0^L \left(z - \frac{L}{2} \right) x^0(z) dz . \qquad (2.57)$$

Therefore,

$$x(z) = x^0(z) + \frac{24(z - L/2)}{L^3} \int_0^L \left(z - \frac{L}{2} \right) x^0(z) dz , \qquad (2.58)$$

and using the mechanical boundary conditions ($\sigma_1 = \sigma_2$, $\sigma_3 = \sigma_4 = \sigma_5 = \sigma_6 = 0$), the total internal elastic energy for each "layer" can then be expressed as:

$$F_{el}(z) = \overline{C} \left[\left(Q_{12}[P^2(z) - <P>^2] + \left(z - \frac{L}{2} \right) \kappa \right) \right]^2 , \qquad (2.59)$$

where $\overline{C}$ is an effective elastic constant given by:

$$\overline{C} = C_{11} + C_{12} - \frac{2C_{12}^2}{C_{11}} , \qquad (2.60)$$

and C_{ij} are the elastic moduli at constant polarization.

4.4   Polarization Profile and Charge Offset

Consider a perovskite ferroelectric oxide such as BaTiO$_3$, which exhibits a cubic-tetragonal ferroelectric phase transformation. Systematic variations in the polarization can be achieved in a number of ways including a variation in the composition of the material, impressing temperature gradients across the structure, or by imposing non-uniform external stress fields, as illustrated in Figure 1.8. For analysis, we assume a ferroelectric of thickness L sandwiched between two metallic electrodes with the easy axis of polarization along the z-axis such that $\mathbf{P}=[0,0,P(z)]$. The ferroelectric is assumed to be homogeneous along the x- and y-directions, reducing the problem to only one dimension. Taking into account the depoling effect and the energy of the internal stresses due to variations in the self-strain (or lattice parameter variations), the Gibbs free energy (per unit area) is given by:

$$
\begin{aligned}
G &= \int_0^L (F_L + F_{el} + F_D)dz \\
&= \int_0^L \left[\frac{1}{2}\alpha P(z)^2 + \frac{1}{4}\beta P(z)^4 + \frac{1}{6}\gamma P(z)^6 + \frac{1}{2}D\left(\frac{dP(z)}{dz}\right)^2 \right. \\
&\qquad \left. - \frac{1}{2}\xi E_D(z)\cdot P(z) + F_{el}(z) \right]dz
\end{aligned}
\tag{2.61}
$$

The above integral can be realized as the sum of the Landau potentials of "layers" with $P(z)$=const., (see Figure 2.12) and their electrostatic and mechanical interaction. The inhomogeneous nature of the three polarization-graded systems is reflected through the position-dependent expansion coefficients with respect to spatial temperature, composition and strain variations.

Neglecting the depoling field, the minimization of the free energy with respect to the polarization in the absence of an external electric field yields the Euler-Lagrange equation:

$$
D\frac{d^2P}{dz^2} = AP + BP^3 + \gamma P^5,
\tag{2.62}
$$

with renormalized coefficients:

$$
A = \alpha + 4\overline{C}Q_{12}\left[\left(z - \frac{L}{2}\right)\kappa - Q_{12} <P>^2\right],
\tag{2.63}
$$

$$
B = \beta + 4\overline{C}Q_{12}^2
$$

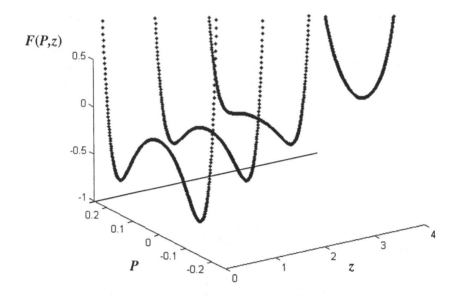

Figure 2.12. The uncoupled Landau potentials of a polarization-graded ferroelectric bar

For compositionally graded ferroelectrics, A, B, and D are a functions of the composition, and therefore are location dependent [i.e., $A(z)$, $B(z)$, and $D(z)$].

For temperature-graded ferroelectrics, A and D are a function of the temperature and are location dependent. Assuming that steady-state heat transfer is established (i.e., the temperature across the ferroelectric bar in Figure 1.8 is a linear function of the position), the normalized coefficient A in Equation (2.63) becomes:

$$A(z) = \frac{L(T_1 - T_0) + z(T_2 - T_1)}{L\varepsilon_0 C}$$
$$+ 4\overline{C}Q_{12}\left[\left(z - \frac{L}{2}\right)\kappa - Q_{12} < P >^2\right] . \tag{2.64}$$

A strain-graded ferroelectric can be analyzed in terms of a simple cantilever beam setup (see Figure 1.8). A bending force is applied along the z-direction resulting in a systematic variation along the z-direction for the normal strain (compressive or tensile). The resulting stress conditions in the cantilever beam are $\sigma_1 \neq 0$ and $\sigma_2 = \sigma_3 = 0$. For a small bending angle θ, the coupling between the polarization and the applied bending force can be neglected yielding the Euler-Lagrange equation:

$$D\frac{d^2P}{dz^2} = (\alpha - 2Q_{12}C^*x_1)P + \beta P^3 + \gamma P^5, \qquad (2.65)$$

where

$$C^* = C_{11} - \frac{2C_{12}^2}{(C_{11} + C_{12})} \qquad (2.66)$$

is an effective modulus for bending and $x_1(z)$ is the position dependent normal strain due to the external bending force. The normal strain $x_1(z)$ is related to the bending angle θ and the length of the bent beam S through:

$$x_1(z) = \left(z - \frac{L}{2}\right)\frac{\theta}{S}. \qquad (2.67)$$

Basic electrostatic theory shows that an inhomogeneous distribution of the polarization is associated with a bound charge (Jackson 1998):

$$\rho_v = -\nabla_i \cdot P_j = -\frac{dP(z)}{dz}, \qquad (2.68)$$

where ρ_v is the volume density of the bound charge. This bound charge generates a built-in electrical field and the resulting built-in potential V_{int} is given by:

$$V_{int} = -\frac{1}{C_F L}\int_0^L z\rho_v(z)dz = \frac{1}{C_F L}\int_0^L z\left(\frac{dP(z)}{dz}\right)dz, \qquad (2.69)$$

where C_F is the capacitance of the graded ferroelectric. Then, the charge offset due to this built-in potential is:

$$\Delta Q = C_Q V_{int} = \frac{k}{L} \int_0^L z \left(\frac{dP(z)}{dz} \right) dz, \qquad (2.70)$$

where C_Q is the capacitance of the load capacitor in a Sawyer-Tower circuit and $k=C_Q/C_F$.

Using the boundary conditions $dP/dz=0$ at $z=0$ and $z=L$ corresponding to complete charge compensation at the ferroelectric/electrode interfaces, we plot the polarization profile normalized with respected to the average polarization $<P>$ for the three cases in Figure 2.13. $P(z)$ is numerically obtained from Equations 2.62-2.70 employing a finite difference method having an accuracy of (successive iterations) approximately 10^{-7}. For the analysis we have chosen BaTiO$_3$ as the prototypical system for the temperature and strain graded ferroelectrics, and Ba$_x$Sr$_{1-x}$TiO$_3$ for the analysis of a compositionally graded ferroelectric system with $0<x<1$. The latter was selected primarily because there exists a great deal of information on the thermodynamic parameters and physical properties of BaTiO$_3$ and SrTiO$_3$. The coefficients α, β, γ and the elastic constants C_{11} and C_{12} for Ba$_x$Sr$_{1-x}$TiO$_3$ were obtained by averaging the corresponding parameters of BaTiO$_3$ and SrTiO$_3$ due to lack of thermodynamic data on single crystals of Ba$_x$Sr$_{1-x}$TiO$_3$. The electrostrictive coefficients are assumed to be insensitive to variations in composition and temperature in the range of the analysis. The thermodynamic and physical properties of BaTiO$_3$ and SrTiO$_3$ used in the calculations are summarized in Table 2.1.

Table 2.1. Thermodynamic and physical properties of Ba$_x$Sr$_{1-x}$TiO$_3$

	BaTiO$_3$	SrTiO$_3$	Ba$_x$Sr$_{1-x}$TiO$_3$
T_0 (°C)	118	-253	$371x$-253
C (°C)	1.7×10^5	0.8×10^5	$(9x+8)\times10^4$
α (m/F)	$6.65\times10^5(T$-118)	$1.41\times10^6(T$+253)	$1.12\times10^7 (T$-$71x$+253)/$(9x+8)$
β (m^5/C^{2}F)	3.56×10^9	8.4×10^9	$(-11.96x$+8.4)$\times10^9$
γ (m^9/C^{4}F)	2.7×10^{11}	-	2.7×10^{11}
C_{11} (N/m^2)	1.76×10^{11}	3.48×10^{11}	$(3.48$-$1.72x)\times10^{11}$
C_{12} (N/m^2)	8.46×10^{10}	1.00×10^{11}	$(1$-$0.154x)\times10^{11}$
Q_{12} (m^4/C^2)	-0.043	-	-0.034

As can be seen from Figure 2.13, the normalized polarization decreases monotonically across the structure with a variation of the polarization along z-

direction predicted in all three graded systems. The polarization gradient diminishes close to the surfaces because of the boundary conditions. For the temperature graded $BaTiO_3$ system, the hot end and cold end are chosen to be at T_C and room temperature (RT=25°C) respectively. If the hot end temperature is higher than T_C, a paraelectric region with no spontaneous polarization will form at this end; a feature important for analyzing the charge offset behavior with respect to the temperature as well. Within the ferroelectric region, the polarization profile exhibits similar behavior as the one shown in Figure 2.13. The end point compositions for compositionally graded systems are $BaTiO_3$ (at $z=0$) and $Ba_{0.75}Sr_{0.25}TiO_3$ (at $z=L$) and a linear relationship between the composition and the position is assumed. It is worth mentioning that the magnitude and the direction of the polarization gradient depend upon the temperature, composition and strain gradients. For example, the direction of the polarization gradient will be reversed if the positions of the hot and cold heat sinks are exchanged. In the strain graded $BaTiO_3$ system, the deflection angle θ of the level arm is taken as 1.5°. A more abrupt polarization gradient should be expected with increasing θ.

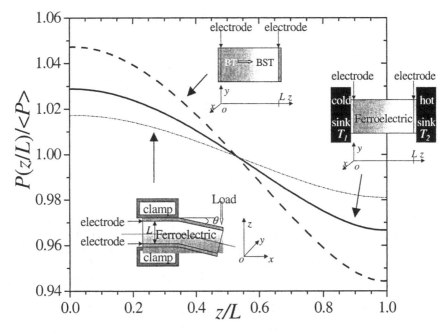

Figure 2.13. Theoretical normalized polarization profiles along the z direction for temperature-, composition-, and strain-graded systems (Ban, Alpay et al. 2003). Reprinted with permission from [Ban, Z.-G., Alpay, S. P. and Mantese, J. V., "Fundamentals of Graded Ferroic Materials and Devices," *Physical Review B* **67**, 184104, (2003)]. Copyright 2003 by the American Physical Society.

In Figure 2.14 we plot the charge offset $\Delta Q/\Delta Q_{max}$ as a function of the normalized parameter ζ for three cases. For temperature graded BaTiO$_3$, ζ is defined as $\zeta = \Delta T/|\Delta T|_{max}$ where $\Delta T = T_2 - T_1$ ($T_1 = RT$ and T_2 is a variable for positive ζ; vice versa for negative ζ). It can be seen from Figure 2.14 that the charge offset first increases in a continuous fashion with increasing ζ, reaching a maximum corresponding to $T_2 = T_0$, and then decreases. Further increase in the temperature leads to the formation of a paraelectric region within the ferroelectric bar with no spontaneous polarization. This region expands with increasing T_2, reducing the charge offset. This very same trend has been observed experimentally in temperature graded Ba$_{0.7}$Sr$_{0.3}$TiO$_3$.

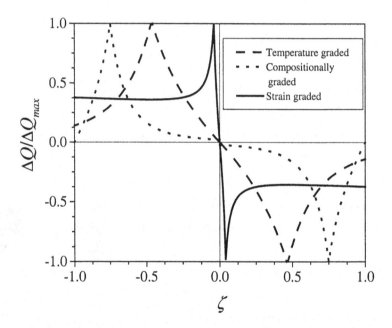

Figure 2.14. Normalized charge offset as a function of the parameter ζ for temperature graded BT, $\zeta = \Delta T/|\Delta T|_{max}$; compositional-graded BST, $\zeta = C_{Ba}$; and strain graded BT, $\zeta = \theta/|\theta|_{max}$ (Ban, Alpay et al. 2003). Reprinted with permission from [Ban, Z.-G., Alpay, S. P. and Mantese, J. V., "Fundamentals of Graded Ferroic Materials and Devices," *Physical Review B* **67**, 184104, (2003)]. Copyright 2003 by the American Physical Society.

A similar behavior is predicted for compositionally graded Ba$_x$Sr$_{1-x}$TiO$_3$ and strain graded BaTiO$_3$. For compositionally graded Ba$_x$Sr$_{1-x}$TiO$_3$, ζ is defined as the Ba concentration on one end of the ferroelectric (i.e., this end has the composition of pure BaTiO$_3$ when $\zeta = 1$ and has the composition of pure SrTiO$_3$ when $\zeta = 0$), and the other end has a fixed concentration corresponding to pure BaTiO$_3$. $\zeta < 0$ corresponds to reversed grading. For strain graded BaTiO$_3$, ζ is the normalized bending angle $\theta/|\theta|_{max}$, where $|\theta|_{max}$ is the maximum bending angle. For compositionally homogenous ($\zeta=1$) or unbent ferroelectric bar ($\zeta=0$), there is no charge offset. A maximum charge offset is predicted at a critical ζ in both cases, corresponding to the emergence of a paraelectric region. However, a steady increase of the charge offset is theoretically expected for relatively larger bending angles in strain graded BaTiO$_3$ due to the saturation of the expansion of the paraelectric region in tensile strain regime and a continuously increasing polarization gradient in the compressive strain regime. An important feature predicted by the model is that the sign of the charge offset is reversed if that of the temperature, composition or strain grading is reversed, as shown in the negative ζ regions in Figure 2.14 justifiable with the same kind of reasoning as discussed for positive ζ.

Although hysteresis offsets have been observed from temperature and strain graded ferroelectric systems, the majority of the research on graded ferroelectrics has concentrated on compositionally graded ferroelectrics. The theoretical modeling for many compositionally graded systems depends on the availability of thermodynamic data and physical properties of single-crystals of not just the end components but in-between solid solutions as well. Furthermore, there are additional factors that have to be taken into account when polycrystalline materials, polycrystalline thin films, and epitaxial thin films are considered.

4.5 Internal Stresses in Graded Thin Films

For thin films (whether they are polycrystalline, textured, or epitaxial) the clamping effect of the substrate, which is usually much thicker than the film, has to be considered. In addition, the coupling of the internal stresses with the self-strain of the phase transformation and thus with the polarization through the electrostrictive effect should also be taken into account. There are several sources of internal stresses in thin films: the structural phase transformation at the Curie temperature, the difference in the thermal expansion coefficients of the film and the substrate, and lattice mismatch between the layers in the case of epitaxial films.

Let us consider a compositionally graded ferroelectric film grown epitaxially on a cubic substrate such that $(001)_{Film}//(001)_{Substrate}$ with the interlayer interfaces being in the xy-plane. We will use the same arguments employed in Sec. 4.3 for a free-standing graded ferroelectric bar and combine it with the stress analysis of Freund (Freund 1996) for compositionally graded semiconductor films.

The mechanical boundary conditions again are given by $\sigma_1 = \sigma_2$ and $\sigma_3 = \sigma_4 = \sigma_5 = \sigma_6 = 0$ for each layer. For a given layer, we define a misfit strain $x_1 = x_2 = x_M$ as:

$$x_M(z) = \frac{a_S - a_0(z)}{a_0(z)} \tag{2.71}$$

where a_S is the lattice parameter if the substrate and a_0 is the cubic lattice parameter of the free-standing film extrapolated to the temperature of the analysis. The component of the self-strain of the ferroelectric phase transformation in the xy-plane is:

$$x^0 = \frac{a(z) - a_0(z)}{a_0(z)} = Q_{12} P(z)^2 \tag{2.72}$$

where a is the constrained lattice parameter of the layer in the ferroelectric phase. Therefore, the total strain for each layer is:

$$x_\Sigma(z) = x_M(z) - x^0(z) + \left(z - \frac{L}{2} \right) \kappa \tag{2.73}$$

where L and κ are the thickness and the radius of curvature of the film, respectively. κ follows from the condition that the average momentum of the internal stress is zero as:

$$\kappa = \frac{24}{L^3} \int_0^L \left(z - \frac{L}{2} \right) [x_M(z) - x^0(z)] dz \tag{2.74}$$

The total strain energy for each layer is given by:

$$F_{El} = \overline{C} x_\Sigma(z)^2 = \overline{C} \left[x_M(z) - Q_{12} P(z)^2 + \left(z - \frac{L}{2} \right) \kappa \right]^2 \tag{2.75}$$

The total thermodynamic potential of an epitaxial graded ferroelectric film is again given by Equation 2.61 with F_{El} of Equation 2.75. The variation in the phase transformation temperature due to misfit (and bending) is due to the renormalization of the Landau coefficient of the P^2-term whereas the clamping effect of the substrate can be understood by the renormalization of the Landau coefficient of the P^4-term.

The epitaxial stresses are partially relaxed by the formation of misfit dislocations at the deposition temperature T_G. The equilibrium thermodynamic theory of misfit dislocations was developed by van der Merwe (Van der Merve 1963) and Matthews and Blakeslee (Matthews and Blakeslee 1974) and is now well established (Nix 1989; Freund 1992). The theory predicts a critical thickness h_ρ below which the formation of misfit dislocations is not feasible. The formation of misfit dislocations to relieve epitaxial stresses in inhomogeneously strained films can be approximated by an "effective" average misfit strain at the deposition temperature that scales with the film thickness h as:

$$< x_M >=< a_0(T_G) > \left(1 - \frac{h_\rho}{h}\right)^{-1} \rho \qquad (2.76)$$

where ρ is the equilibrium *average* linear misfit dislocation density at T_G. Assuming no additional dislocations form during cooling down [i.e., $\rho(T_G)=\rho(T)$ for $T<T_G$], an "effective" substrate lattice parameter, can be defined and used for the calculation of the position dependent misfit strain of the film:

$$\bar{a}_S(T) = \frac{a_S(T)}{\rho\, a_S(T) + 1}. \qquad (2.77)$$

The effect of thermal expansion difference between the graded film and the substrate can be incorporated into the analysis by a careful study of the history of the deposition process. The total strain at the ambient temperature will be related to the deposition temperature as well as the degree of relaxation provided by misfit dislocations at the deposition temperature. For textured and polycrystalline graded ferroelectric films, the misfit strain in Equation 2.75 has to be replaced by the thermal strain due to the difference between the film and the substrate given by:

$$x_T(z) = [\lambda_{Film}(z) - \lambda_{Substrate}]\Delta T \qquad (2.76)$$

where λ_{Film} and $\lambda_{Substrate}$ are the thermal expansion coefficients of the film and the substrate, respectively, and ΔT is the difference between the deposition temperature and the temperature of analysis. It should be noted that for the analysis of polycrystalline graded ferroelectrics, the polycrystalline (isotropic) electrostrictive coefficients and elastic moduli have to be used.

4.6   Dielectric and Pyroelectric Response

The dielectric response of compositionally graded ferroelectrics as a function of temperature exhibits characteristics of a diffuse phase transformation (Cross 1987; Cross 1994), which is inherently linked, with the distribution of the phase transformation temperature resulting from the composition gradient across the ferroelectric. Calculating the polarization profiles with and without a very small applied field δE by using the Euler-Lagrange equation (Equation 2.62):

$$D\frac{d^2 P(z)}{dz^2} = AP(z) + BP(z)^3 + \gamma P(z)^5 - \delta E , \qquad (2.77)$$

we can obtain the relative small-signal dielectric permittivity profile as:

$$\frac{\varepsilon(z)}{\varepsilon_0} = \frac{1}{\varepsilon_0}\frac{P(z, E = 0) - P_2(z, E = \delta E)}{\delta E} = \frac{1}{\varepsilon_0}\frac{\Delta P(z)}{\delta E} . \qquad (2.78)$$

Figure 2.15 shows the polarization profile and the corresponding dielectric permittivity profile for temperature-graded BaTiO$_3$. The temperatures at the two ends of the ferroelectric are chosen to be at T_C and room temperature 25°C respectively. The increment of electric field is taken as $\sim 10^{-5}$ kV/cm. The dielectric permittivity increases monotonically along the z-direction, consistent with the fact that there is a continuous spatial variation of polarization across the structure. This shows that the polarization grading gives rise to a non-homogeneous distribution of the dielectric permittivity in the materials. The region having larger polarization exhibits smaller dielectric response and a nearly invariable permittivity is predicted close to the surface as a result of the diminishing polarization gradient.

The average small-signal permittivity $<\varepsilon_m>$ of the ferroelectric bar is given by (Scaife 1998):

$$\frac{<\varepsilon_m>}{\varepsilon_0} = L\left[\int_0^L \frac{\varepsilon_0}{\varepsilon(z)}dz\right]^{-1} . \qquad (2.79)$$

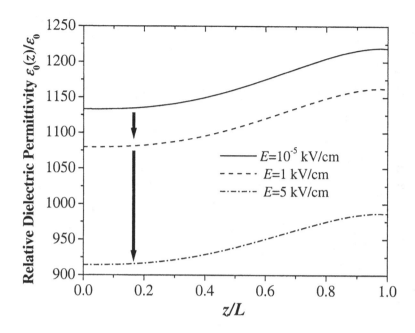

Figure 2.15. Dielectric permittivity profile as a function of the external field E for compositionally graded $Ba_{0.75}Sr_{0.25}TiO_3$.

Figure 2.16 shows the variation in the average dielectric response as a function of temperature for three compositionally graded $Ba_xSr_{1-x}TiO_3$ systems with one end fixed at $BaTiO_3$. In comparison to a sharp peak of the dielectric permittivity at T_C for bulk homogenous ferroelectrics, a typical diffused dielectric response with the temperature is predicted for the compositionally graded $Ba_xSr_{1-x}TiO_3$ ferroelectrics. The maximum in the dielectric permittivity is broadened and the extent of this broadening depends on the composition gradient. A steeper composition gradient can give rise to a broader maximum. The exact same behavior has been documented experimentally in compositionally graded $Ba_xSr_{1-x}TiO_3$ ferroelectric thin films where a more pronounced broad plateau region of the permittivity with the variation of the temperature was observed for $Ba_{0.5}Sr_{0.5}TiO_3$–$BaTiO_3$ graded thin film compared to $Ba_{0.75}Sr_{0.25}TiO_3$–$BaTiO_3$ graded film (Slowak, Hoffmann et al. 1999). Another important aspect of the theoretically calculated dielectric permittivity is that the broad plateau is shifted towards lower temperatures with an increase in the compositional gradient. This is understandable because the "average" Curie temperature of the graded ferroelectric decreases with increasing compositional gradient. Similar results should be expected for

temperature and strain-graded ferroelectrics where the "average" Curie
temperature is a function of the strength of the temperature or strain gradient.

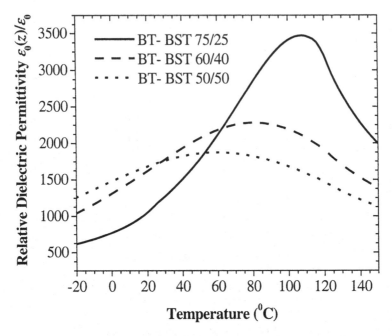

Figure 2.16. The relative average dielectric permittivity as a function of *T* for a compositionally
graded BT (Ban, Alpay et al. 2003). Copyright 2003 from "Hysteresis Offset and Dielectric
Response of Compositionally Graded Ferroelectric Materials," *Integrated Ferroelectrics*, 58:
1281-1291 by Ban, Z. –G. et al. Reproduced by permission of Taylor & Francis, Inc.,
http://www.taylorandfrancis.com.

The conventional pyroelectric coefficient *p* is defined in Equation 2.13.
For polarization-graded ferroelectrics, the hysteresis loop is shifted "up" or
"down" along the polarization axis, resulting in a charge offset. In this case,
an effective (or pseudo-) pyroelectric coefficient p_{eff} can been defined as:

$$p_{eff} = \frac{d(\Delta Q)}{dT}.$$ (2.80)

Substituting for ΔQ (Equation 2.70), we obtain

$$p_{eff} = \frac{k}{L}\frac{d}{dT}\int_0^L z\left(\frac{dP}{dz}\right)dz.$$ (2.81)

Figure 2.17 shows the theoretically calculated temperature dependence of the charge offset and the effective pyroelectric coefficient for compositionally graded $Ba_xSr_{1-x}TiO_3$. The end compositions are chosen to be $BaTiO_3$ and $Ba_{0.7}Sr_{0.3}TiO_3$ respectively. A maximum effective pyroelectric coefficient is expected to occur around 15°C, corresponding to the onset of paraelectricity at the $Ba_{0.7}Sr_{0.3}TiO_3$ end. The effect of the strength of the composition grading is shown in Figure 2.18 where we plot the effective pyroelectric coefficient at room temperature as a function of barium composition for compositionally graded $Ba_xSr_{1-x}TiO_3$ ($Ba_xSr_{1-x}TiO_3–BaTiO_3$). The maximum pyroresponse is for $Ba_{0.75}Sr_{0.25}TiO_3–BaTiO_3$, since the transformation temperature for $Ba_{0.75}Sr_{0.25}TiO_3$ is around room temperature. Steeper gradients with barium composition at one end less than 0.75 will result in paraelectric regions in the graded ferroelectric that will lower the pyroresponse.

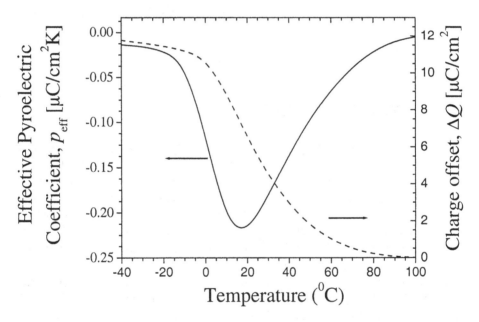

Figure 2.17. The effective pyroelectric coefficient (solid line) and charge offset (dashed line) as a function of temperature for compositionally graded $BaTiO_3$-$Ba_{0.7}Sr_{0.3}TiO_3$.

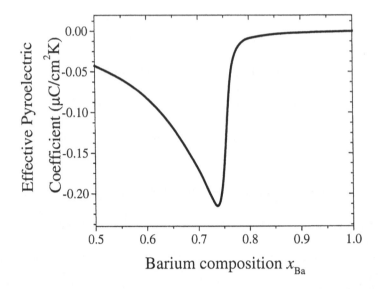

Figure 2.18. The effective pyroelectric coefficient as a function of Ba composition for compositionally graded $Ba_xSr_{1-x}TiO_3$.

5.   POLARIZATION DYNAMICS

Although there remains much to be understood in order to describe theoretically the dynamic properties of polarization-graded ferroelectrics, we will attempt to provide an initial analysis based on the thermodynamic approach developed in this Chapter and basic statistical mechanics. It is clear that the switching properties and the asymmetric shift of the polarization hysteresis along the polarization axis (the charge offset) is a result of domain phenomena. Should domains form due to a steep polarization gradient, the domain structure of polarization-graded ferroelectrics is expected to be significantly different than the $180°$-domains in homogeneous ferroelectrics (Figure 2.9). Clearly, further experimental work is needed to elucidate this. Although a single domain state throughout the sample can be stabilized for smooth gradients, the reversal of polarization by an external electric field should proceed via domain phenomena. It is not our intention to propose a domain configuration that minimizes the depoling field or a scenario for domain nucleation and growth in graded ferroelectrics at this time. Instead, we will concentrate on the driving force for the polarization offset. In a way, this is similar to the thermodynamic analysis of the electric field dependence of the polarization in a homogeneous ferroelectric where the polarization reversal is the result of the instability of one ferroelectric ground state with respect to the other in the presence of a critical (coercive) field. It is well documented that such an approach yields coercive fields that are two to three orders of magnitude larger than the experimentally measured values. Therefore, it is clear that the analysis provided in this Section for graded ferroelectrics will suffer from the very same shortcomings of the application of the thermodynamic model to homogeneous ferroelectrics. However, such an approach would be a starting point as a qualitative description of the field dependence of the polarization and stimulate further theoretical and experimental research.

Polarization-graded ferroelectrics can be thought of a series of slightly skewed double-well potentials aligned along the z-axis. The depth of the wells and hence the height of the potential barrier varies with position along the chemically/thermally/mechanically-graded axis as shown schematically in Figure 1.8. This implies that there is a lower energy state that is preferred for each well pair commensurate with a chemical, thermal, or mechanical gradient. Statistically, this implies that with all things being equal, there is a slightly higher probability for the lower well to be populated more than the upper well at each well pair.

Let us first demonstrate that the free energy functional is indeed tilted towards one ferroelectric ground state by re-visiting the analysis of Sec. 4.2 of a stress-free ferroelectric bilayer. The total free energy (per unit area) of the coupled bilayer structure shown in Figure 2.10 is given by:

$$F_\Sigma = (1 - \alpha)F_1 + \alpha F_2 + F_S / h \qquad (2.82)$$

where F_S is the interfacial energy per unit area of interfaces. The surface energy of interfaces reflects the additional energy contribution to the bilayer system due to compositional inhomogeneity and polarization discontinuity, which extend over a very small length scale. For the sake of simplicity, we assume that layer 1 and 2 fit perfectly to each other such that there is no elastic energy and no electrostrictive coupling between layers. We will further assume that our bilayer is at a temperature sufficiently away from the transformation temperature of either layer and the gradient terms in either layer due to thermal fluctuations are thus negligible. The free energies of layers 1 and 2 (per unit area) are:

$$F_1 = F_{0,1} + \frac{1}{2}aP_1^2 + \frac{1}{4}bP_1^4 - \frac{1}{2}E_{D,1}P_1 \qquad (2.83)$$

$$F_2 = F_{0,2} + \frac{1}{2}cP_2^2 + \frac{1}{4}dP_2^4 - \frac{1}{2}E_{D,2}P_2 \qquad (2.84)$$

where $E_{D,i}$ are the depoling fields given in Sec. 4.2. It is clear from the above Landau functional that the phase transformation characteristics of the bilayer are significantly different than their bulk constituents. The depolarization term in the free energy relations can be considered as a "symmetry-breaking" field that results in the preference of one ferroelectric ground state over the other (compared to two equivalent but oppositely polarized ferroelectric ground states in a homogeneous material). We will address this in more detail later in this Section.

Combining terms, we obtain for the total energy:

$$F_\Sigma = (1-\alpha)\left[F_{0,1} + \frac{1}{2}a*P_1^2 + \frac{1}{4}bP_1^4\right] + \alpha\left[F_{0,2} + \frac{1}{2}c*P_2^2 + \frac{1}{4}dP_2^4\right] \quad (2.85)$$
$$+ JP_1P_2 + F_S / h$$

where $a*$ and $c*$ are re-normalized Landau coefficients given by:

$$a* = a + \frac{\alpha\xi}{\varepsilon_0}, \quad c* = c + \frac{(1-\alpha)\xi}{\varepsilon_0}, \quad (2.86)$$

and J is a coupling coefficient given by:

$$J = -\frac{\alpha(1-\alpha)}{\varepsilon_0}\xi \quad (2.87)$$

that describes the strength of the electrical interaction between layers. The interaction term JP_1P_2 in the above free energy functional is in complete agreement with TIM models and other Landau models [see for example (Qu, Zhong et al. 1997; Ma, Shen et al. 2000)].

The electrical interaction manifests itself via the re-normalization of the dielectric stiffness of the two layers due to the depolarization effect. This interaction alters all phase transformation characteristics as well as the electrical and electromechanical properties of the individual layers. The variation in the Curie temperatures of layer 1 and 2 follow from $a*=0$ and $c*=0$, and are given by:

$$\Delta T_{C,1} = -\alpha\xi C_1 \quad (2.88)$$
$$\Delta T_{C,2} = -(1-\alpha)\xi C_2 \quad (2.89)$$

The equilibrium polarization in the two layers can then be calculated using simultaneous solution of the equations of state:

$$\frac{\partial F_\Sigma}{\partial P_1} = (1 - \alpha)(a * P_1 + b P_1^3) + J P_2 = 0,$$
(2.90)

$$\frac{\partial F_\Sigma}{\partial P_2} = \alpha(c * P_2 + d P_2^3) + J P_1 = 0.$$
(2.91)

As it was shown in Section 4.2, a single-domain state can be stabilized if the initial uncoupled polarization difference between layers is small or the bound charge due to the polarization jump at the interlayer interface is compensated by the flow of free charges within the bilayer. In Figure 2.19a and b, we plot the equilibrium polarization of both layers as a function of their volume fraction and ξ which is related to the density of free charges for a BaTiO$_3$-Ba$_{0.99}$Sr$_{0.01}$TiO$_3$ bilayer. It can be seen that the single-domain state is stable for $0 < \alpha < 1$. The spontaneous polarization difference between the layers is smaller when compared to their 1.8% difference in bulk. For smaller ξ (see Figure 2.19b), the difference between the polarizations between layers may be larger since some of the bound charge at the interlayer interface may be compensated with free charge carriers.

More interestingly, in Figure 2.20 we plot the free energy as a function of the polarization in both layers that clearly shows the symmetry-breaking effect of the built-in potential due to the bound charge at the interface. All the free energy curves are normalized with respect to $F_{0,i}$. At $\xi = 0$, both layers exhibit a typical symmetric double well potential with the minima corresponding to two energetically identical ferroelectric ground states with $|P| > 0$. Electrical coupling arises at non-zero ξ values and is enhanced with increasing ξ such that the symmetric double wells in both layers are now skewed towards one ferroelectric state with $P > 0$. For a perfect insulating bilayer ($\xi = 1$), it can be seen that there is maximum electrical interaction between layers that results in the stability of just one ferroelectric ground state in both layers albeit a small polarization difference. This shows that polarization graded ferroelectrics with a smooth gradient and insulating properties may very well be self-poled. These results are entirely consistent with the Slater model developed by Mantese *et al.* (Mantese, Schubring et al. 1997) for compositionally graded ferroelectrics and recent first principles calculations for ferroelectric perovskites with compositional inversion symmetry breaking (Sai, Meyer et al. 2000).

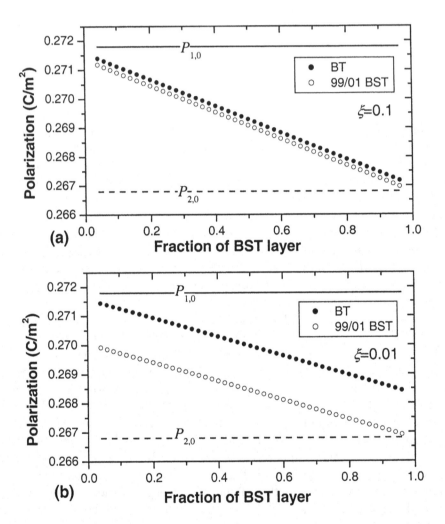

Figure 2.19. Equilibrium spontaneous polarization in $BaTiO_3$ (BT)–$Ba_{0.99}Sr_{0.01}TiO_3$ (99/01 BST) bilayer as a function of free charge density ξ. The bulk polarization of BT and 99/01 BST also shown.

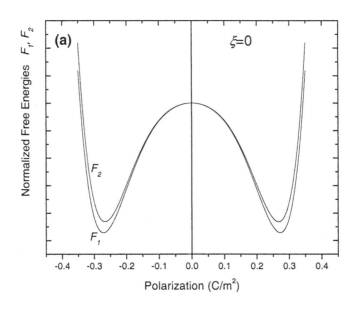

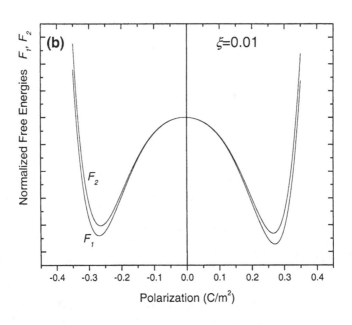

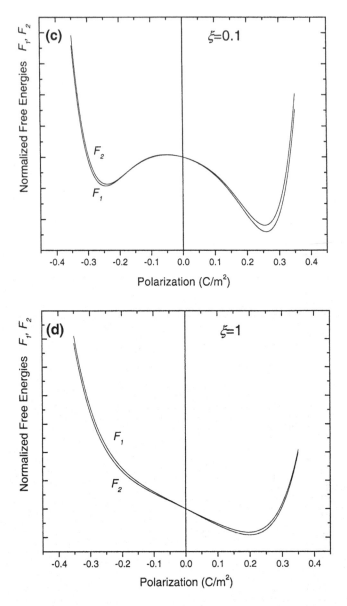

Figure 2.20. Free energy potentials for an equi-fraction BaTiO$_3$ (BT)–Ba$_{0.99}$Sr$_{0.01}$TiO$_3$ (BST 99/01) bilayer. (a)ξ=0 corresponding to semiconductor ferroelectrics, layers are electrically uncoupled, (b) ξ=0.01, (c) ξ=0.1, (d) ξ=1, corresponding to insulating ferroelectrics resulting in self-poling in both layers.

One would be tempted to take our thermodynamic analysis a bit further to try and understand the effect of an applied field. Due to the polarization mismatch at the interlayer interface, no matter how small it may be, there exists a bound charge. Although the voltage drop across a layer is constant, it follows from Gauss' law that this charge should result in a non-linear distribution of the electric field in both layers that reaches a maximum at the interface and quickly decays on either side of it (Jackson 1998). Furthermore, since both materials are ferroelectric, the relative dielectric constant will be both position and electric field dependent. Their relation, of course, is non-linear. This necessitates finding a way to self-consistently solve for the electric field in the bilayer as a function of position, field-dependent polarizations, and permittivities.

A qualitative treatment of the switching properties of graded ferroelectrics description based on statistical mechanics was given in Chapter 1. In Figure 2.21b we show schematically the skewed potential that follows from the analysis for a bilayer at a given z. For comparison, Figure 2.21a plots the potential if the ferroelectric were homogeneous with a composition (or equivalently at a given temperature or mechanical strain) at position z. At a given temperature, the probability of "jumps" from one ferroelectric ground state in the homogeneous material is given by $exp(-U/k_BT)$ where k_B is the Boltzmann constant and U is the energy barrier which is essentially the difference between the energies in the ferroelectric and paraelectric states, $F(P=P_S)-F(P=0)$ or equivalently $F(P=-P_S)-F(P=0)$. Thus, it is clear for the homogeneous case (Figure 2.21a), that both ferroelectric states are equally occupied and an applied field would tip this balance towards the ferroelectric state with the favorably oriented polarization direction with respect to the field. The applied field would decrease the potential barrier towards one preferred orientation of the polarization and thus the probability of occupation in that ferroelectric ground state would increase.

For the case of a built-in potential the symmetry is broken and one ferroelectric state is preferred over the other even in the absence of an external field. When an electric field is applied, the barrier decreases even further. At a large enough field, the height of the wells at all z will be very small or non-existent which is when the maximum polarization offset is reached. Obviously, the polarization reversal is not homogeneous as the wells at each and every z is non-symmetric. Therefore, the polarization offset follows the direction of the polarization grading.

The above analysis is consistent with experimental findings. Indeed, experimentally one sees that a polarization-graded ferroelectric responds much like a hysteresis. However, as the field is increased slightly above zero a steady offset along the polarization axis reveals itself. The offset is often found to increase with according to a power law of the form, $P_S \propto (E_{max})^\gamma$, with $\gamma \sim 2$, where E_{max} is the maximum of the periodic external field (often sinusoidal excitation). The offsets continue to increase with the applied field until either the material arcs through resulting in catastrophic device failure, or consistent with our model at the highest field, again diminish to zero. In the latter case, the applied field overwhelms the internal potentials of the devices.

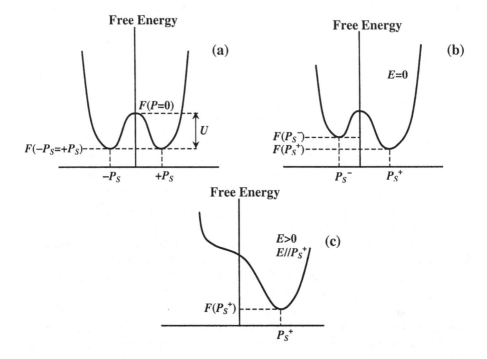

Figure 2.21. Schematic free energy potentials as a function of polarization for (a) homogeneous ferroelectric with two identical but oppositely oriented polarization directions, (b) an "upgraded" ferroelectric at a position z, (c) same as (b) with an applied electric field along P_S^+.

Chapter 3

GRADED FERROELECTRIC DEVICES (GFDS)

1.   INTRODUCTION

Our discussion of the fabrication and characterization of polarization-graded devices begins with the materials from which they are formed. One seeks material systems in which the local polarization density, $\mathbf{P(r)}$, can be specified through the material chemistry alone; analogous to the selection of specific semiconductor materials (e.g. silicon or gallium arsenide) which can be locally doped with either donor or acceptor species to form *p-n* or *n-p* junction devices. In hindsight, ferroelectrics are the obvious materials of choice from which to construct GFDs; for below a critical temperature (the Curie temperature, T_C) they spontaneously polarize, yielding a net dipole moment density, $\mathbf{P(r)} \neq 0$, which is fixed by the material composition.

As described in Chapter 1, the most conspicuous signature of a polarization-graded ferroelectric is an asymmetry in its *Q-V* characteristics. The internal potential of a GFD is revealed as a translation along the polarization axis, when the device is placed in an otherwise ordinary Sawyer-Tower circuit (Sawyer and Tower 1930). We, therefore, present such results as an indication of the presence of a GFD structure; however, more detailed analyses (such as *I-V* measurements and chemical depth profiling) are required to ensure that one is not observing experimental or system artifacts. We address this latter topic below.

2.   FERROELECTRIC MATERIAL SYSTEMS

Ferroelectric materials are actually quite common. In general, one can identify them by their non-centrosymmetric charge distributions which give rise to spontaneous dipole moments. Ferroelectrics, include representatives from the perovskites, tungsten-bronzes, Aurivillius compounds, halides, nitrites, and sulfates. Many of the more well known ferroelectrics have been catalogued by Lines and Glass (Lines and Glass 1977).

The most extensively studied ferroelectric materials are the perovskites; particularly at temperatures about their cubic to tetragonal phase transformation. Ferroelectric perovskites also usually have tetragonal to orthorhombic and orthorhombic to rhombohedral transitions at temperatures lower than the cubic to tetragonal transition (Lines and Glass 1977). In the discussion that follows, we focus only on the primary transition of this material system.

The general chemical representation for materials belonging to the perovskite family is ABO_3, where A is a monovalent, divalent or trivalent metal and B is a pentavalent, tetravalent or trivalent element. The size of the A and B ions are such that they yield a "tolerance" factor close to unity, forming the basic perovskite structure shown in Figure 2.1 a and b.

Barium titanate, $BaTiO_3$, is the prototypical ferroelectric perovskite with a cubic to tetragonal transition of approximately 120°C. The variation in the spontaneous polarization in $BaTiO_3$ as a function of temperature is shown in Figure 3.1 (Jona and Shirane 1962). The actual Curie temperatures of ferroelectric perovskites are functions of many factors, including impurities, internal stress, and electric field as noted in Chapter 1; a fact we will exploit later in this chapter. Strontium titanate, $SrTiO_3$, forms a solid solution with barium titanate, to yield $Ba_{1-x}Sr_xTiO_3$ (BST) which have Curie temperatures which decrease with increasing $SrTiO_3$ content, see Figure 3.2a (Hilton and Ricketts 1996). Likewise, the spontaneous polarization is found to decrease approximately linearly with increasing $SrTiO_3$ content, Figure 3.2b (Hilton and Ricketts 1996). This latter feature of BST will be especially important for the formation of GFDs. Barium titanate and its variants are used throughout the electronics industry for applications requiring surface mount or on-chip capacitors (Kaiser March 1993).

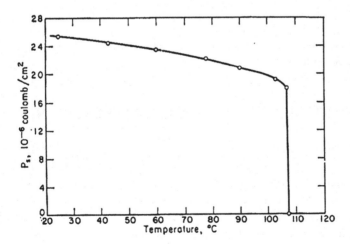

Figure 3.1. Plot of the polarization versus temperature for $BaTiO_3$ (Jona and Shirane 1962).

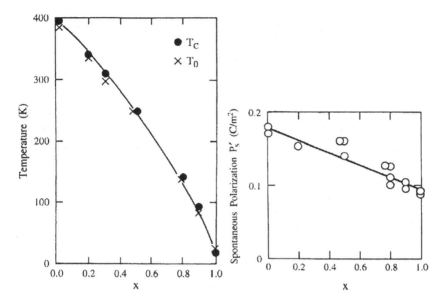

Figure 3.2. (a) Variation in the Curie temperature of BST as a function of SrTiO$_3$ content. (b) Spontaneous polarization for bulk BST as a function of SrTiO$_3$ content (Hilton and Ricketts 1996).

Another intensively studied mixed perovskite is lead zirconate titanate, PbZr$_{1-x}$Ti$_x$O$_3$ (PZT), and its lanthanum modified counterpart, Pb$_{1-3y/2}$La$_y$Zr$_{1-x}$Ti$_x$O$_3$ (PLZT); wherein the polarization is greater for those compositions having either greater Zr or La content (Haun, Furman et al. 1989). These materials have Curie temperatures in excess of 200°C and find extensive use as the base materials for electronically controlled actuators (Haun, Furman et al. 1989; Haun, Furman et al. 1989; Haun, Furman et al. 1989; Haun, Furman et al. 1989; Haun, Furman et al. 1989).

Finally, one often sees in the literature on graded ferroelectrics a number of references to the perovskite, potassium tantalum niobate, KTa$_{1-x}$Nb$_x$O$_3$ (KTN) (Lines and Glass 1977). This material system also has both a Curie temperature and spontaneous polarization that may be varied by tailoring the relative ratio of Ta to Nb, the tantalum rich compounds having the lower Curie temperatures. The work on KTN GFDs, however, is now only of historical note as this material system was that in which the first graded ferroelectrics were formed and from whence many of the preliminary concepts relative to GFDs were first conceived (Schubring, Mantese et al. 1992). Potassium tantalum niobate is currently no longer a preferred material system from which to construct polarization-graded ferroelectrics as its properties are more difficult to control compared to BST, PZT, or PLZT.

3. GROWTH OF GRADED FERROELECTRIC STRUCTURES

When forming graded ferroelectric devices, one usually endeavors to tailor the local chemistry of the material to create uniform gradients in polarization density. The primary means for experimentally realizing these compositional variations have included: solution chemistry methods such as metallo-organic decomposition (MOD) and sol-gel methodologies (Schubring, Mantese et al. 1992; Mantese, Schubring et al. 1997), pulsed laser deposition (PLD)(Brazier, McElfresh et al. 1998), sputtering (Jin, Auner et al. 1998), and molecular beam epitaxy (Tsurumi, Miyasou et al. 1998). Of these techniques, solution chemistry and PLD have been employed most often for creating GFDs; hence we will review that work in the greatest detail.

The very first polarization-graded ferroelectrics were fabricated using MOD (Mantese, Micheli et al. 1989). The working materials were KTN and BST; with polarization density gradients achieved by varying the Ta to Nb or the Ba to Sr ratio (respectively) normal to the growth surface (Schubring, Mantese et al. 1992; Mantese, Schubring et al. 1997).

For solution-based thin film processing, it is necessary to begin with the organic precursors from which the graded structures will be derived. For example, when using the MOD method for the deposition of BST, carboxylates (specifically, the 2-ethyl hexanoates, the oxtoates, or the neodecanoates) of barium and strontium are dissolved in xylene (or other suitable solvent) together with titanium 2-ethylhexoxide. The concentrations of the individual organic compounds are adjusted to yield solutions having net viscosities of approximately 1 cp. Pyrolysis of the BST solutions yield $Ba_{1-x}Sr_xTiO_3$ thin films having the desired Ba to Sr ratio and are stochiometric in titanium. Solution adjustments are most often done using x-ray photospectroscopy (XPS) or energy dispersive x-ray analysis.

Platinum foil or silicon substrates that have been oxidized, metallized with a thin film of titanium (~40 nm), then metallized with much thicker platinum (~300 nm) are often chosen as the substrates on which to grow the compositionally graded thin film oxides. The organics are spun-cast on pre-cleaned substrates at 1000-6000 rpm, which distributes them uniformly over the substrate and drives off the solvent. Subsequently, the organic film is converted to oxides by placing the film on substrate in an air or oxygen muffle furnace (set between 500 and 600°C) for 1-2 minutes. Film thickness may be increased by successive spin-castings and pyrolysis. Upon final pyrolysis, an annealing is usually done in air or oxygen to crystallize the materials and smooth out any compositional abruptness. The annealing treatment may be done at temperatures exceeding 1000°C, depending upon the base material system. A detailed description of the MOD process may be found in the literature (Mantese, Micheli et al. 1989).

To test the concepts outlined in Chapters 1 and 2, compositionally graded BST films were formed by Mantese *et al.* on platinum foil by MOD (Mantese, Schubring et al. 1997). Four separate organic precursor solutions were prepared as described above having Ba to Sr ratios in the pyrolized films of 1:0, 0.9:0.1, 0.8:0.2 and 0.7:0.3. Two layers of each composition were used to create a four-step graded device as shown in Figure 3.3a and 3.3b. Final annealings of the approximately 1-μm thick films were done in oxygen at 1100°C. X-ray photospectroscopy depth profiling revealed a Ba:Sr ratio which varied linearly with depth normal to the growth surface of both devices (Mantese, Schubring et al. 1997).

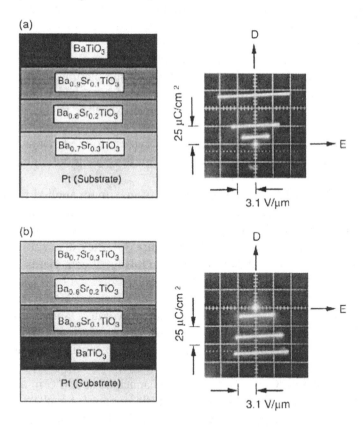

Figure 3.3. Schematic of the BST layered structures prior to annealing to form the GFD's. (a) "Up" and (b) "down" devices. Also shown are the unconventional hysteresis phenomena. Note that in these plots the gain of the amplifier has been decreased to capture the considerable translation of the hysteresis graphs along the displacement axis (Mantese, Schubring et al. 1997).

Capacitor-like structures were formed by the subsequent deposition of gold-chromium, yielding a contact area $A=0.2$ cm^2, contacts through a shadow mask. The gold was approximately 300 nm thick on a 60 nm chromium layer. The platinum substrate served as the counter-electrode. Hysteresis measurements were performed under vacuum at room temperature in a temperature controlled system (stability 0.05°C) using a modified Sawyer-Tower circuit with a 1 kHz sinusoidal voltage (Sawyer and Tower 1930).

In contrast to the usual hysteresis phenomenon of "ungraded" ferroelectric materials (in a normal Q-V plot); Figures 3.3a and 3.3b show (respectively) that the authors observed quite distinct "up" and "down" translations, $\Delta Q = \Delta P_{offset} \cdot A$ (where ΔP_{offset} is the polarization translation), of the hysteresis loops along the polarization axis. The displacements increased with increasing applied voltage. In these hysteresis graphs the gain of the polarization voltage amplifier was greatly reduced so as to capture the substantial translational effect, hence the loops appear as flattened lines in the plots. "Up" and "down" are relative terms in that the platinum substrate was the reference electrode in both hysteresis measurements, hence the sense of the displacements could be reversed by a simple exchange in the position of the reference electrode. The authors did not observe unconventional hysteresis in homogeneous $Ba_{1-x}Sr_xTiO_3$ capacitor devices formed from uniform composition films for the entire range of starting material compositions (Mantese, Schubring et al. 1997).

In Figure 3.4, P_S is plotted as a function of maximum applied field, E_{max}. To obtain P_S as a function of excitation field, Mantese *et al.* ac coupled the inputs to the scope used in a Sawyer-Tower circuit. A best fit to the data indicates that $P_S \propto (E_{max})^\gamma$, with $\gamma \sim 2.4$. In Figure 3.4, they also show that the measured field dependence of ΔQ. A best fit to this latter data yields, $\Delta Q \sim E^{4.4}$.

Similar results were also obtained for graded $Pb_{1-x}La_xTi_{1-x/4}O_3$ thin films deposited on $LaNiO_3$ coated silicon/silicon oxide substrates (Bao, Mizutani et al. 2000). For these GFDs lead acetate trihydrate, lanthanum nitrate hexahydrate, and titanium tetra-n-butoxide were used as the starting precursors with methoxyethanol as the solvent. Three separate solutions with the nominal compositions of $x=0$, 12, and 25 mole percent were spun-cast, pyrolyzed in air at 550 °C for 60 minutes, and annealed at 650°C in air.

Figures 3.5a and 3.5b show the hysteresis curves for both "upgraded" ($x=0$, 12, 25) and "downgraded" ($x=25$, 12, 0) GFDs. In Bao *et al.*'s work, they observed $\Delta Q \sim E^{3.02}$ for their upgraded films (Bao, Mizutani et al. 2000). They also presented the *I-V* characteristics for this same film, Figure 3.6, showing symmetric positive and negative bias characteristics with near zero leakage. These same authors also reported similar findings for graded ferroelectric structures that were formed from calcium-modified lead titanate $Pb_{1-x}Ca_xTiO_3$, with power-law characteristics for both P_s and ΔQ (Bao, Mizutani et al. 2001).

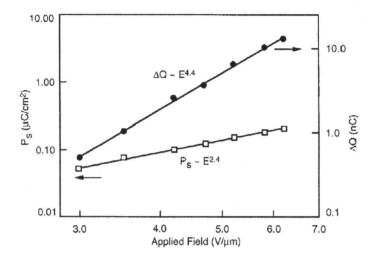

Figure 3.4. Plot of the spontaneous polarization as a function of maximum field for a 1 μm-thick graded $Ba_{1-x}Sr_xTiO_3$ film. Also shown is the variation in the charge offset, ΔQ, as a function of field for the same film (Mantese, Schubring et al. 1997).

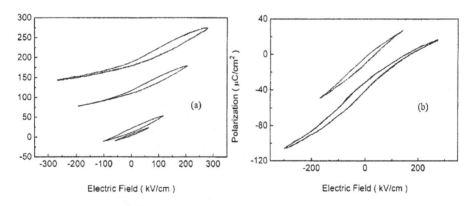

Figure 3.5. Hysteresis graphs of (a) "up" (x = 0, 12, 25) and (b) "down" (x = 25, 12, 0) $Pb_{1-x}La_xTi_{1-x/4}O_3$ GFD devices (Bao, Mizutani et al. 2000).

Pulsed laser and molecular beam epitaxy thin film deposition methods have offered the greatest control in the precise formation of graded ferroelectric structures. In particular, the PLD work by Brazier *et al.*, yielded "up" and "down" graded $PbZr_{1-x}Ti_xO_3$ (PZT), films having a near continuous gradient in Zr:Ti ratio (Brazier, McElfresh et al. 1998). To form these structures, the authors used a split laser target having compositions of $PbZr_{0.75}Ti_{0.25}O_3$ and $PbZr_{0.55}Ti_{0.45}O_3$ respectively, see Figure 3.7; then rastered the laser along the compositional break line. The approximately 0.3 μm thick compositionally graded films were deposited upon platinum foil with 50 μm x 50 μm platinum top electrodes; thereby forming symmetric electrical contact structures. Figure 3.8 shows an Auger depth compositional profile of the Zr to Ti content for a typical "down-graded" film.

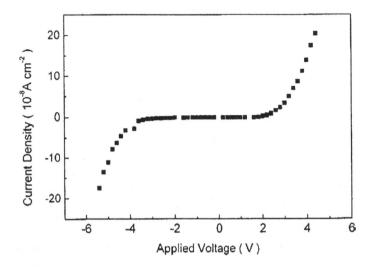

Figure 3.6. Current density versus applied voltage for the "upgraded" film shown in Figure 3.5a (Bao, Mizutani et al. 2000).

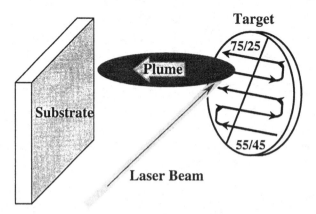

Figure 3.7. Method used by Brazier et al to grow continuously variable composition $PbZr_{1-x}Ti_xO_3$ (PZT) films (Brazier, McElfresh et al. 1998).

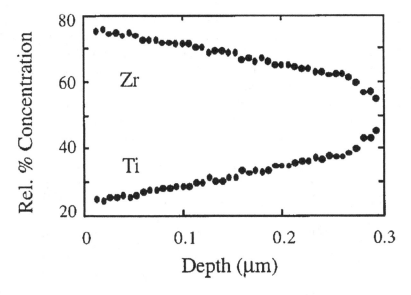

Figure 3.8. Zr to Ti ratio for a compositionally graded $PbZr_{1-x}Ti_xO_3$ film (Brazier, McElfresh et al. 1998).

The resultant hysteresis graphs for "up-graded" (Zr:Ti ratio varied from 75:25 to 55:45, highest Zr content at the platinum substrate) and "down-graded" (Zr:Ti ratio varied from 55:45 to 75:25, lowest Zr content at the platinum substrate) $PbZr_{1-x}Ti_xO_3$ films are shown in Figure 3.9a and 3.9b. Likewise, Figure 3.10 is a plot of ΔQ versus applied field, showing an approximate E^5 dependence. The authors noted *"extremely low dc leakage currents (10 GΩ resistance)"* with symmetric *I-V* characteristics.

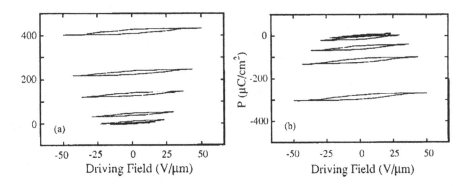

Figure 3.9. Hysteresis graphs of (a) "upgraded" (x = 0.25 to 0.45) and (b) "down" (x = 0.45 to 0.25) $PbZr_{1-x}Ti_xO_3$ GFD's devices (Brazier, McElfresh et al. 1998).

The power law dependence of ΔQ has been observed by a number of groups in a variety of material systems (Mantese, Schubring et al. 1997; Brazier, McElfresh et al. 1998; Bao, Mizutani et al. 2000). While a heuristic argument has been given to justify such a functional dependence, neither thermodynamic nor *ab initio* models have been developed sufficiently to adequately explain these observations. Indeed, an understanding of the time and field-strength dependence of ΔQ remains one of the outstanding problems pertaining to polarization-graded structures and devices.

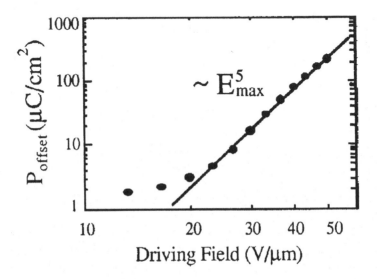

Figure 3.10. Typical field dependence of the hysteresis offsets for $PbZr_{1-x}Ti_xO_3$ GFD's (Brazier, McElfresh et al. 1998).

4.   OTHER POLARIZATION-GRADED FERROELECTRICS

Graded ferroelectric structures and devices have been fabricated using a variety of materials, employing a number of deposition methods, with similar experimental results observed by many research groups. However, alternative interpretations of the experimental observations, other than the one presented in Chapter 1, have been proposed to account for the aberrant hysteretic behavior of GFDs; including, charge injection at the electrodes, non-uniformities in the thin film structures due to asymmetries in the growth process, leakage currents, and electrical breakdown of the films (Brazier and

McElfresh 1999; Boerasu, Pintilie et al. 2000; Poullain, Bouregba et al. 2002; Chan, Lam et al. 2004).

Recall, however, in Chapter 1, Figure 1.8, it was argued that GFDs could be formed by any process which results in a polarization gradient normal to the measurement axis of the device; including, not only GFDs which arise from compositional gradients, but those whose origin stem from either temperature or stress gradients.

Therefore, in two fundamental experiments, one done with bulk ceramics and the other done using stress-graded thin films, compelling evidence was assembled legitimizing the above observations as arising from the wholly new phenomena of polarization-graded ferroelectrics. Hence, the origin of the offsets can now unequivocally be related to spatial polarization variations in the ferroelectric materials.

5.   TEMPERATURE-GRADIENT INDUCED GFDS

In their work on compositionally homogeneous bulk BST ceramics, Fellberg *et al.* used standard ceramic processing methods to prepare the samples tested, including: wet mixing of precision weights of barium titanate and strontium titanate fine-grained reagent powders, isostatic compaction, sintering to near theoretical density with no open porosity, wafering, and polishing the materials to shape (Fellberg, Mantese et al. 2001). The samples were in the form of ~0.1 cm thick ceramic disks, approximately 1 cm in diameter and metallized on both sides by evaporation of 60 nm of chromium followed by 300 nm of gold. In all cases the composition of the ferroelectric was chemically uniform throughout the samples.

Fellberg *et al.* adjusted the composition of the ferroelectric starting materials to have a barium titanate ($BaTiO_3$) to strontium titanate ($SrTiO_3$) ratio of 75:25 yielding a material with a Curie temperature, see Figure 3.11, of approximately 45°C. In these figures the relative permittivity, normalized to vacuum, is plotted as a function of temperature.

In Figure 3.12 the spontaneous and remnant polarization are plotted as a function of temperature. In the inset of Figure 3.12 the authors show an example of hysteresis (an electric displacement versus applied field plot, $D=\varepsilon_o E+P$), taken at 30°C. In the absence of a temperature gradient, Fellberg *et al.* noted that the ferroelectric samples yielded only normal hysteresis with no offset, "up" or "down", over the entire temperature range reported. Hence, she concluded that artifacts due to rectifying contacts, asymmetries in composition, charge injection, electrical leakage or breakdown, etc., were non-existent in her samples and could be excluded as a cause for the results observed in the presence of a temperature gradient. Note that the polarization had a maximum at 35°C, and then decreased rapidly with temperature.

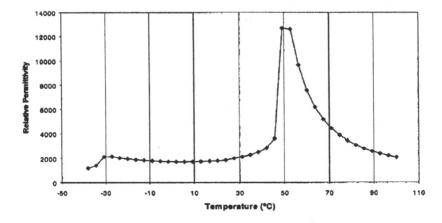

Figure 3.11. Plot of the relative permittivity, normalized to vacuum, as a function of temperature for a bulk ferroelectric ceramic (0.1 cm thick disk) having a barium titanate to strontium titanate ratio of 75:25. The line is a polynomial fit to the data (Fellberg, Mantese et al. 2001).

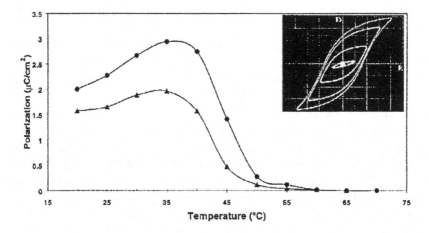

Figure 3.12. Plot of the spontaneous (●) and remnant (▲) polarization as a function of temperature. The lines are polynomial fits to the data. Normal hysteresis in the absence of a temperature gradient when the sample was held at 30 °C (Fellberg, Mantese et al. 2001).

Prior to Fellberg *et al.*'s work, it had been postulated (Mantese, Schubring et al. 1995), but never before been demonstrated, that "up" and "down" offsets could occur in bulk materials by imposing a temperature gradient across a ferroelectric material near its Curie temperature. According to previous arguments (Mantese, Schubring et al. 1995; Mantese, Schubring et al. 1997), hysteresis offsets could arise from temperature induced polarization gradients within the material because the polarization is such a strong function of temperature near the Curie point. This is illustrated schematically in Figure 3.13, which shows how the polarization would be effected in a ferroelectric material with a temperature difference impressed across the sample.

Cold Copper   Hot Copper
Heat Sink   Heat Sink

Figure 3.13. Schematic of a temperature induced polarization density gradient for a bulk ceramic $Ba_{1-x}Sr_xTiO_3$ with one end held at 35 °C and the other at a temperature slightly greater (Fellberg, Mantese et al. 2001).

In Figure 3.14, Fellberg shows a schematic of her test setup to measure hysteresis in the presence of a temperature gradient. A Sawyer-Tower circuit (Sawyer and Tower 1930) was used with a 100 V peak sine wave at an excitation frequency of 1 kHz. The position of the ground reference electrode was kept constant throughout the complete set of measurements, which was the bottom electrode in the discussion below. The difference in temperature across the 0.1 cm thick sample, ΔT, was such that at least one of the copper heat sinks was always fixed at 35°C. A positive ΔT, therefore, corresponded to heating the top heat sink to >35°C, while a negative ΔT corresponded to heating the bottom heat sink to >35°C. Thus, the temperature gradient that was impressed across the sample produced the greatest variation in polarization across the sample.

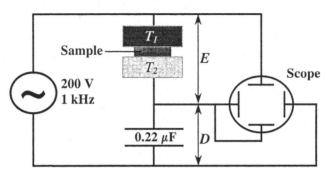

Figure 3.14. Experimental setup to measure hysteresis in the presence of a temperature gradient (Fellberg, Mantese et al. 2001).

In Figure 3.15 the authors clearly show that for $\Delta T=0$, no offset were observed. For small ΔT, where the polarization changes slowly with temperature, the hysteresis offset was virtually indistinguishable from the background noise; whereas, the offsets were easily observed where the polarization varies rapidly with temperature; i.e., near the material's Curie temperature. Moreover, the offset changed sign with a change in the sense of the temperature gradient. The offsets disappeared when the material ceased to be ferroelectric.

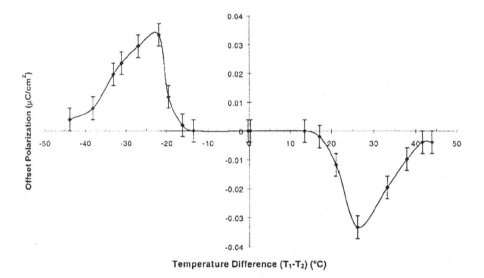

Figure 3.15. Plot of the hysteresis offset, reported as an offset polarization, as a function of temperature difference between the two heat sinks (Fellberg, Mantese et al. 2001).

To understand the results in Figure 3.15 and to connect to the discussion in Chapter 2 we consider a mono-domain ferroelectric of thickness L, sandwiched between two metallic electrodes as illustrated in Figure 3.16 (Alpay, Ban et al. 2003). The two electrodes are in contact with thermal sinks at temperatures T_1 and T_2, imposing a temperature gradient across the ferroelectric bar. The easy axis of polarization is along the z-axis such that $\mathbf{P}=[0,0,P(z)]$ and the ferroelectric is assumed to be chemically and thermally homogeneous along the x- and y-directions reducing the problem to only one dimension. The LGD free energy (per unit area) is given by Equation 2.61. Assuming that the "global" depoling field is diminished due to the formation of electrical domains and approximating the Ginzburg coefficient A as $\delta^2 |\alpha|$, where δ is the characteristic length along which the polarization varies, we arrive at the Euler-Lagrange equation (Equation 2.62) with the renormalized

Landau coefficients given by Equations 2.64 under steady-state heat transfer conditions.

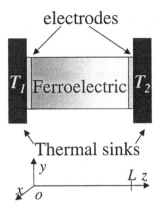

Figure 3.16. Schematic diagram showing the setup of the temperature graded ferroelectric and the coordinate used in the calculation.

Using the boundary conditions $dP/dz=0$ at $z=0$ and $z=L$ corresponding to charge compensation at the ferroelectric/electrode interfaces, we plot the normalized polarization with respect to average polarization $<P>$ as a function of position for three different temperature gradients for BST 75/25 in Figure 3.17 (T_1 is fixed at 35^0C if $T_2/T_1>1$; or T_2 is fixed at 35^0C if $T_2/T_1<1$). $P(z)$ is numerically obtained from Equations 2.61, 2.62, and 2.64 using a finite difference method having an accuracy of (successive iterations) approximately 10^{-7}. As can be seen from Figure 3.17, the normalized polarization varies monotonically along the z-direction. The polarization gradient diminishes close to the surfaces because of the boundary conditions. The magnitude and the direction of the polarization gradient depend on the temperature gradient imposed on the ferroelectric. If $T_2/T_1>1$, i.e., there is a positive temperature gradient along the z-direction, a continuously declining polarization profile is obtained. The polarization profile becomes steeper if the temperature gradient is increased from $T_2/T_1=1.3$ (solid line) to $T_2/T_1=1.6$ (dashed line). The dash-dotted line in Figure 3.17 corresponds to the case where a negative temperature gradient ($T_2/T_1=0.6$) is imposed on the ferroelectric bar. In this case, the polarization profile is inverted in comparison to the positive temperature gradient.

The polarization profile can then be used to calculate the charge offset per unit area ΔQ by using Equation (2.68-2.70). To evaluate the effect of the temperature gradient on the charge offset, we fix the temperature T_1 as a constant ($T_1=35^{\circ}$C) and vary the temperature T_2 ($T_2 \geq T_1$). In Figure 3.18, we plot the charge offsets $\Delta Q/\Delta Q_{max}$ as a function of the temperature difference $\Delta T = T_2 - T_1$ for temperature-graded BST 75/25 (solid lines). Figure 3.18 shows that the charge offset initially increases in a continuous fashion with increasing temperature difference ΔT as a result of the increasing polarization gradient. A maximum charge offset is obtained when hot sink temperature T_2 reaches a temperature where the polarization is varying most rapidly, though the materials has not converted to the paraelectric phase. Further increase in the temperature leads to the formation of a paraelectric region within the ferroelectric bar with no spontaneous polarization. This paraelectric region expands with increasing T_2, resulting in a decrease in the charge offset. An important feature predicted by the model is that the sign of the charge offset is reversed if that of the temperature grading is reversed (i.e., T_2 is fixed at 35°C; T_1 is a variable and $T_1 \geq T_2$), as shown in the negative temperature difference region in Figures 3.18.

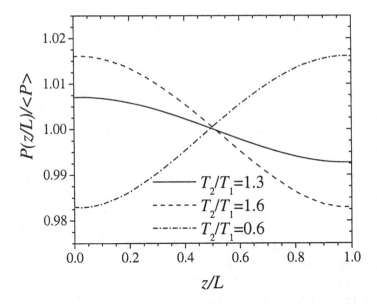

Figure 3.17. Theoretical, normalized polarization profiles along the z-direction for temperature-graded $Ba_{0.75}Sr_{0.25}TiO_3$. T_1 and T_2 are the temperatures at each end of the ferroelectric bar. The dash-dotted line corresponds to the case where a negative temperature gradient ($T_2/T_1=0.6$) imposed on the ferroelectric bar (Alpay, Ban et al. 2003).

The above theoretical prediction of the charge offset for temperature-graded ferroelectrics is compared to the experimental results, as shown by the squares in Figure 3.18. The agreement between the experimental data and theoretical calculation is quite remarkable considering that there are no adjustable parameters in the theoretical analysis. It should be noted that for small temperature gradients, the experimentally determined charge offsets are indistinguishable from the background variation. The small difference between the theoretical and experimental results may be explained by considering the clamping effect of the electrodes by the heat sinks. In the experimental measurement, the whole sample was under compression so as to maintain good thermal contact between the heat sinks. Consequently, as temperature gradients were established across the sample through heating, the sample expanded; resulting in an increase in the thermal stress, which diminished the overall polarization. The corresponding charge offset is, therefore, suppressed for very small temperature gradients. As it is quite difficult to accurately model the combined electro-mechanical setup without introducing a number of adjustable free parameters, we have not tried to directly account for this effect.

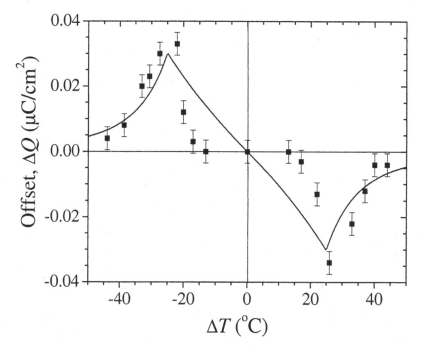

Figure 3.18. Calculated charge offset as a function of the temperature difference ΔT between hot and cold sinks for temperature graded $Ba_{0.75}Sr_{0.25}TiO_3$ (solid lines) (Alpay, Ban et al. 2003).

The above results are quite compelling in establishing the existence and origin of polarization-graded ferroelectrics. For if one argued that other artifacts were the cause of the observed offsets in bulk ceramics, these extrinsic sources would still be present in the absence of the temperature gradient. In addition, the observation of both types of hysteresis offsets, "up" and "down", in the bulk materials, directly links the offsets with the establishment of a polarization gradient across the sample.

6.   STRAIN-GRADIENT INDUCED GFDS

Because the ferroelectric spontaneous polarization, P_S, is a function of: material composition, c, temperature, T, and stress, σ, i.e., $P_S = P_S(c, T, \sigma)$; it has been possible to form GFDs from a variety of material systems, both by grading the composition of the ferroelectric and by imposing temperature gradients normal to the electrode surfaces (Mantese, Schubring et al. 1995; Mantese, Schubring et al. 1997). Moreover, as just discussed, these findings are fully consistent with theoretical analysis. Recent experimental research, however, has gone yet further in creating GFDs by the imposition of stress/strain gradients on homogeneous materials. Specifically, in a series of experiments on strain-graded ferroelectric films, Mantese *et al.* not only demonstrated the formation of GFDs by imposing stress gradients across their thin film material, but also were able to alter the resultant polarization gradients through variation of the magnitude of the gradient (Mantese, Schubring et al. 2002).

In their work, lead strontium titanate (PST) thin films, ~ 4 μm thick, were deposited on 0.02 cm thick platinum foil substrates by MOD. In particular, metal oxide precursors consisting of: titanium (IV) 2-ethylhexoxide, lead (II) neodecanoate, and strontium neodeconoate were combined together in proper proportions, then diluted with xylene to yield a single precursor solution. Films of $Pb_{0.4}Sr_{0.6}TiO_3$ – PST (with a composition determined by X-ray photospectroscopy) were deposited onto platinum substrates through spin coating and pyrolysis then annealed at 950 °C for 60 minutes in air. Scanning electron microscopy revealed that the films had grains which were ~0.2-0.3 μm in diameter. X-ray diffraction analysis clearly indicated a perovskite crystal structure (tetragonal) at room temperature.

From substrate bow measurement it was determined that the PST films (as formed) were under compression (see Figure 3.19). The radius of curvature was determined from the bow profile to be 2.61 m with a maximum displacement of ~17 μm along the film-substrate normal. The film compression arises from the difference in thermal expansion coefficients of the film and the substrate (11.8×10^{-6} °C^{-1} and ~11×10^{-6} °C, for Pt and PST, respectively) as the film is cooled down from the annealing temperature. Therefore, as formed, the PST films had a built-in stress gradient normal to the platinum substrate, with a stress maximum at the film surface (Figure 3.19).

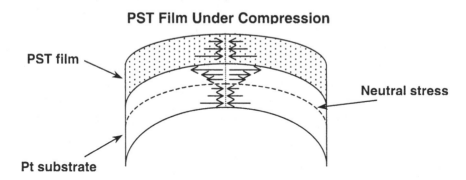

Figure 3.19. As formed, the PST films are in compression with a stress gradient normal to the platinum substrate (Mantese, Schubring et al. 2002).

Gold/chromium electrodes were deposited upon the PST surface in the form of circular dots (~ 0.02 cm^2) using electron beam evaporation. The platinum substrate served as the counter and reference electrode contact when the devices were inserted into a Sawyer-Tower circuit (Sawyer and Tower 1930). All such characterization was done at room temperature with 10 V peak, 10 kHz sine wave excitation. The low voltage PST "capacitance" was ~3 nF. A 50 nF sampling capacitor was used in a Sawyer-Tower set-up.

The PST system is remarkably similar to the BST system in that the Curie temperature for the cubic to tetragonal transition can be adjusted over an exceedingly large temperature range by adjusting the lead to strontium ratio. With a Pb:Sr ratio of 4:6 the Curie temperature of the PST thin film material is slightly below room temperature. Figure 3.20a is a plot the relative permittivity (normalized to vacuum) as a function of temperature. Note the broad peak centered around 15°C. Full conversion from the tetragonal to the cubic phase (with increasing temperature) is not abrupt, and Mantese *et al.* found that measurable hysteresis persisted to nearly 150°C, see Figure 3.20b (Mantese, Schubring et al. 2002).

To form GFDs from the homogeneous PST films, a gradient in the polarization was established between the electrode surfaces through the imposition of a stress gradient normal to the growth surface of the film. To accomplish this task in a well-defined geometry, the substrate and brass thermal heat sink were sandwiched between two electrically insulated blocks with the bend point along the diameter of one of the gold/chromium electrical contacts as shown in Figure 3.21a. Because the platinum substrate was very much thicker than the PST film, the neutral stress line for the bi-layer should lie within the platinum when a bend angle θ was imposed. Consequently, the PST as a whole remains in compression (only for small θ); though, superimposed on this compressional stress field there was a gradient in stress (either tensile or compressional) due to the externally applied force as shown in Figure 3.21b. To obtain a reliable measure of θ, the platinum substrate extending beyond the clamped region was sandwiched between two electrical quality fiberboards to serve as an 82.5 mm lever arm. The end-deflection of the lever arm was measured using a modified micrometer mechanism capable of $\pm$ 0.1 mm resolution. Theta, the deviation angle, could thus be measured as the arctangent of the deflection/lever arm ratio with a $\pm$ 0.07° uncertainty.

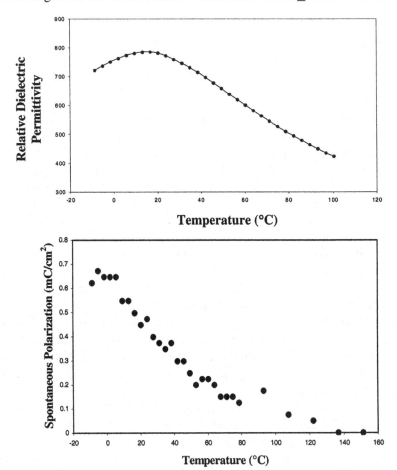

Figure 3.20. (a) Relative permittivity (normalized to vacuum) as a function of temperature for a ~ 4 μm thick $Pb_{0.4}Sr_{0.6}TiO_3$ – PST film deposited on a platinum substrate. (b) Spontaneous polarization as a function of temperature for the same PST film (Mantese, Schubring et al. 2002).

Because the measured hysteresis offsets were quite small, the traditional Sawyer-Tower circuit required modification to reliably obtain reproducible results with high precision and low uncertainty. In particular, the voltage across the PST film was measured using a dual input Tektronix 7A22 differential amplifier plug in unit into a Tektronix 7904 oscilloscope, so that only the voltage across the sample was recorded. The plus and minus spontaneous polarization values, as measured across the series 50 nF capacitor, were limited to three significant digits with a net difference between the two readings limited to $\pm$ 1mV uncertainty. Thus the very stable and reproducible offset measurements were accurate to within $\pm$ 0.00024 $\mu C/cm^2$.

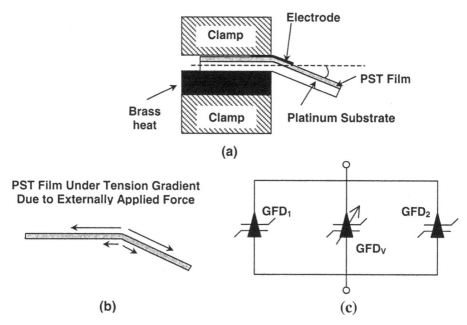

Figure 3.21. (a) Experimental arrangement for imposing a stress gradient in the central region of the PST capacitor structure. Note that θ is negative as depicted (*externally applied* <u>*tension*</u>). (b) Stress distribution due to bending action only, ignoring built in compression. (c) Equivalent circuit of (b) (Mantese, Schubring et al. 2002).

While the authors had some concern that thermal drift would obscure the effects due to the imposed stress gradients, they concluded that the placement of the brass heat sink in contact with the platinum substrate was sufficient to ensure the stability of the measurement against thermal drift. Indeed, the stability of the system was such that no measurable deviation from a recorded offset would occur during an over-night measurement hiatus.

In Figure 3.22 Mantese plots the polarization displacements, ΔP, as a function of bend angle, θ. Some features of this figure are worthy of special note: (1) There was an observable, positive, net offset even in the absence of an externally imposed stress gradient, i.e., $\theta{=}0$. (2) The variation in offset was nearly symmetric for positive and negative θ, though film compression (positive θ) had less effect on the observed offset than when tension (negative θ) is externally applied to the film. (3) The offset, ΔP, changes by only ~ $\pm$ 10% as the bend angle varied from -10° to +10°.

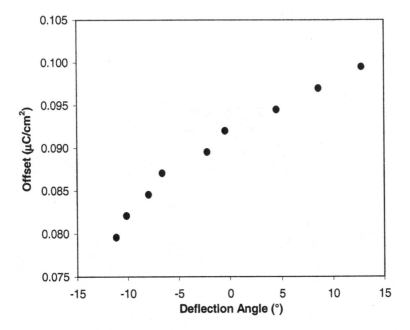

Figure 3.22. Hysteresis offset, ΔP, as a function of bend angle θ (Mantese, Schubring et al. 2002).

To understand these results, recall that substrate bow measurements indicated that, as formed, the PST films were under compression. Thus, a stress gradient existed across the film even in the absence of an externally applied stress gradient (see Figure 3.19). Therefore, it is not surprising that a polarization offset was observed even for the unbent $\theta{=}0$ state. This result is consistent with previous work on barium strontium titanate films deposited by molecular beam epitaxy upon single crystal strontium titanate (Tsurumi, Miyasou et al. 1998). There it was found that the presence of a substrate-imposed stress gradient was the primary factor in determining the direction and magnitude of the polarization offset; the presence of a compositional gradient in Ba:Sr ratio normal to the substrate only modified the primary effect.

To complete our analysis of the results of Figure 3.22 it is important to understand the experimental setup as depicted in Figure 3.21a. Recall, that the film/substrate bi-layer was clamped along a line that traversed the diagonal of the gold/chromium top electrode. Thus, even though the entire electrode area was quite small (~0.020 cm^2) the stress-changed area (arising from θ) of the PST capacitor was much smaller than the entire sample area – as represented by the electrode area. The applied stress gradient experiment can thus be depicted schematically as shown in Figure 3.21c. Here, GFD$_1$ and GFD$_2$ represent the GFDs that were formed due to the compressional stress gradient across the PST film imposed by the difference in expansion of the platinum and PST ceramic when the perovskite phase of the PST is formed. These GFDs remained essentially unchanged as θ was varied. GFD$_V$, however, represents a variable GFD who's stress gradient varied with bend angle, θ. Of course, by construction, the three GFDs are electrically in parallel.

Our understanding of Figure 3.22 now becomes clearer from the above discussion. One would: (1) Expect a hysteresis offset for θ=0 due to the compressional stress in the as-formed PST films. (2) Expect some degree of symmetry for $\pm\theta$ as the "built-in" and externally imposed stress gradients are mostly uncoupled. (3) Expect the externally imposed stress gradient to produce a small variation in ΔP due to the fact that the θ effected zone of the PST sample is only a small fraction of the overall electrode area.

7.   CONCLUDING REMARKS

It has often been (correctly) stated that the offsets observed in the hysteresis graphs of polarization-graded ferroelectrics can be duplicated in homogenous ferroelectric materials with the use of diodes, capacitors, and other suitable silicon circuitry; a "Silicon Equivalent" (SE) analysis. Thus, it is concluded, the characteristics of GFDs are of extrinsic origin (Brazier and McElfresh 1999; Boerasu, Pintilie et al. 2000; Poullain, Bouregba et al. 2002; Chan, Lam et al. 2004). The rationale behind this line of reasoning is that the internal construct of GFDs lead to rectifying junctions or other silicon-like components at the electrode-ferroelectric interfaces or within the structure of the devices themselves and therefore do not represent the intrinsic properties of polarization-graded ferroelectrics.

While SE constructs can most assuredly duplicate the response of GFDs, they cannot reasonably account for the polarization offsets of single crystal GFDs or bulk ceramics. For GFDs formed by means of temperature gradients, especially, one would expect the silicon-like elements to reveal themselves in the absence of the temperature gradients, which is not the case. Moreover, the authors of the studies on GFDs often took great pains to examine the *I-V* characteristics of the devices to rule out conduction current asymmetries. Indeed, in one case, self-supporting polarization-graded ferroelectrics were grown with a variety of symmetric metallizations, all yielding similar results (Schubring, Mantese et al. 1992). In addition, one must ultimately reconcile SE analysis with Equation 1.6; which is based solely upon Maxwell's relations. The inescapable conclusion of Equation 1.6 is that it requires the existence of an internal potential if a polarization gradient is present within the material. One can only circumvent this requirement by assuming ferroelectric material conductivities far in excess of what one traditionally observes in high quality thin film ferroelectric materials.

Finally, if SE arguments are to be the ultimate arbiter of device physics, one would likewise conclude from Figure 3.23 that the hysteretic responses of uniform polarization ferroelectric capacitors are themselves artifacts of the electrodes and dielectrics. For, one can adjust the relative values of the silicon-based components to realize almost any ferroelectric characteristic. Surely, such a conclusion would not be consistent with the accepted physics of these materials (Sheikholeslami and Gulak 1997; Rep and Prins 1999).

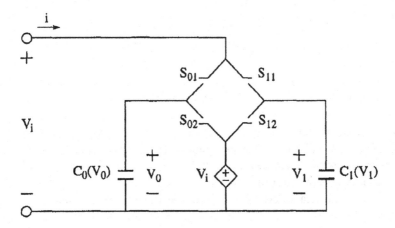

Figure 3.23. All-silicon equivalent circuit of a ferroelectric capacitor (Sheikholeslami and Gulak 1997). © 1997 IEEE.

Chapter 4

TRANS-CAPACITIVE DEVICES: TRANSPACITORS

1. INTRODUCTION

Passive semiconductor components such as diodes, Zener's, thermistors, LED's, and varactors have a myriad of uses; including applications for: electromagnetic filtering, rectifiers, sensors, and displays. The greatest impact, however, of spatially non-homogeneous semiconductor structures has been from active elements, with transistors being foremost in creating the most significant technological changes.

From the outset, the inventors of the transistor envisioned that their device would be capable of power gain; i.e., have the ability to amplify a small signal energy source by a near-linear process. Indeed, one of the very first transistor experiments was to demonstrate that it could be used as an audio amplifier. Acting much like an electronic switch, transistors regulate the flow of current from an energy reservoir through a resistive load. A relatively small input current acts as the metering element. With this control scheme, even simple transistor devices are capable of delivering signal power gains of 10,000. For transistors, the DC power supply voltages act as the reservoirs from which energy may be extracted and sourced to the circuit loads.

It was likewise anticipated by the researchers who studied GFDs that the most interesting polarization-graded ferroelectrics would be those excited by oscillatory external source potentials, from which energy could be metered out and stored by passive loads (Mantese, Schubring et al. 2001; Mantese, Schubring et al. 2002). Gating of these energy reservoirs was similarly found to be accomplished using small energy inputs to the polarization-graded structures. As in previous chapters, we follow an historical account of the description and fabrication of trans-capacitive device elements: the active counterparts of transistors.

2.   TRANS-RESISTIVE ELEMENTS
         (TRANSISTORS): POWER GAIN DEVICES

Transistors can be thought of most simply as continuously variable resistive devices that are capable of being electronically switched from fully on to fully off. The current flow through a transistor is controlled via its base current that acts much like a gate in a water-flow circuit under a pressure head. The transistor itself, however, does not generate energy. Its primary function is to meter out energy from external sources in proportion to the control signal current into its base.

We shall not delve into the structure of the many types of transistor devices, nor is it necessary to discuss their detailed band structures to understand their operation. Instead, we shall suffice it to say that amplification occurs through "transistor action", which is an all inclusive term we will use to describe the overall properties of transistors in the absence of a detailed analysis; the latter of which can be found in a variety of texts devoted to the subject (Ashcroft and Mermin 1976; Streetman 1980; Sze 1981).

One of the most common transistor configurations is the *emitter-follower* shown in Figure 4.1. For small (assuming the device has been biased correctly) incremental current inputs to its base, i_b, one observes an emitter current i_e, into the load, R, equal to:

$$i_e \approx h_{fe} i_b \qquad\qquad (4.1)$$

h_{fe} is referred to as the transistor beta and typically is on the order of 100. Thus, an *emitter-follower* is a current amplifier, with a current gain approximately equal to the beta of the transistor.

Transistor action therefore enables the signal power to be amplified by a gain factor of:

$$G \equiv (h_{fe} i_b)^2 / (i_b)^2 = h_{fe}^2 \approx 10,000 \qquad\qquad (4.2)$$

Additional benefits of an emitter follower also include: high input impedance and low output impedance; essential amplifier characteristics when attempting to use a high output source impedance to drive a low impedance load.

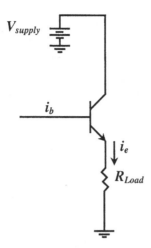

Figure 4.1. n-p-n transistor in an *emitter-follower* configuration.

The key property of a transistor in the emitter-follower configuration is its ability to produce power gain. While other passive devices are capable of producing either voltage or current gain, e.g. transformers, they do not amplify power. It is this single aspect that is primarily responsible for the transistor's great utility.

Of course, there are many other transistor configurations of equal or greater importance (depending upon the application) that have been exploited for both digital and analogue signal processing. Indeed, far too many to even attempt to enumerate. However, we wish to briefly discuss one other configuration, the *push-pull follower* designed to handle inputs with both positive and negative signal excursions.

Emitter-followers require biasing in order to avoid transistor cutoff for input signals having both negative and positive current/voltage excursions. Consequently, it is often necessary to strip off the bias voltage in follow-on amplifier elements or loads as they can drive elements into non-linear response. The *push-pull* configuration is a two-transistor circuit, see Figure 4.2. In operation, T_1 conducts during positive signal swings, while T_2 is active for negative swings; thereby permitting unbiased signal amplification. Such a configuration (and it variants) finds uses in audio circuits; however, it too has limitations that make it undesirable for high fidelity applications.

Finally, the implementation and integration of transistor technology has been further facilitated by the development of simple hybrid circuit models (e.g., the Ebers-Moll parameter models) that greatly assist the circuit designer at the board level (Ebers and Moll 1954). Likewise, SUPREME, PICES, ANSYS, SPICE and many other computer aided design software programs have enabled IC manufacturers the ability to design their semiconductor junction devices with near exact performance characteristics.

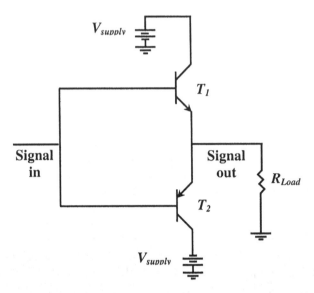

Figure 4.2. *n-p-n* and *p-n-p* transistors configured as a *push-pull* amplifier.

Without a doubt, transistor technology and history is incredibly rich and represents the summation of many man-years of effort. The primary purpose of this short overview; however, is not to cover the complete history and development of transistor technology, but to allow us to draw parallels to this device and to potential applications of its dielectric counterpart.

3.   TRANSPACITORS

To construct a device that may be viewed as the dielectric analogue of the transistor (e.g. in the *emitter-follower* configuration), we must have a number of elements that are analogous to this transistor configuration: a signal input or source, an energy supply, a load, and a means of amplifying signal power. By referring to Table 1.1, insights can be gained toward this construction.

We confine ourselves to dipolar materials with few free charge carriers, which suggests that our loads must be capacitive in nature and our energy sources be AC potential excitations instead of DC. Thus, in Figure 4.3 we have replaced the traditional DC potential and resistive load with an AC supply and capacitive load.

In the case of a transistor, amplification is realized by altering the internal potential of a *p-n-p* or *n-p-n* structure through the application of an external input thereby permitting free charge to flow through the structure. For the dielectric case, we desire that the internal potential likewise be a function of an external stimulus. To capture this aspect of a dielectric amplifier we recall that the internal potential of a GFD is a function of: composition grade, temperature, and stress. Thus, we imagine the signal input of the device to be an external energy source, principally: temperature, voltage, or strain as depicted in Figure 4.3.

Finally, we imagine, that the trans-capacitive circuit configuration must have some element of asymmetry to it, primarily through the construction of the elements that make up the amplifier. This feature is realized by selecting an "up" or "down" GFD as the working element.

Transpacitor

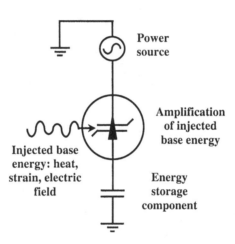

Figure 4.3. Dielectric analogue of an *emitter-follower* amplifier.

Figure 4.3 thus depicts one construct of a trans-capacitive amplifier. The astute reader, however, may object that our analogy with the transistor is imperfect, as we do not have opposing "up" and "down" GFDs as might be expected from a direct comparison with the internal structure of a transistor. Indeed, transistors are often depicted as opposing diodes as shown in Figure 4.4. The reason for this imperfect analogy resides in the fact that a transistor's base materials are equally capable of conducting either DC or AC currents. Hence, in a transistor, to affect the internal potential which allows amplification; we must forward bias (for an *n-p-n* device) the base-emitter diode and reverse bias the collector-base diode. For trans-capacitive devices, however, DC is blocked by the GFD internal "capacitance" (strictly speaking, a GFD has a field-dependent permittivity and so the term varactor is more appropriate). Therefore, the internal potential of a dielectric amplifier may be altered through the direct application of a DC potential. Here we assume that the load capacitor is very much greater than the capacitance of the GFD. We relax this requirement in the *push-pull* configuration described later in this chapter.

Recalling that an *emitter-follower* is best described as a current amplifier (Equation 4.1), and referring back to Table 1.1, we anticipate that the trans-capacitive configuration of Figure 4.3 should be describable as a charge amplifier. Indeed, such is the case. To test this concept, KTN "up" devices were inserted into the structure shown in Figure 4.3. A 1 kHz sine wave potential with peak amplitude of 10 V excited the circuit. The load to GFD capacitance ratio was 220.

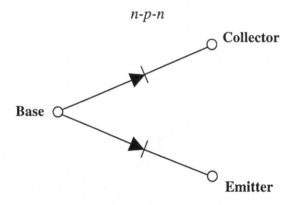

Figure 4.4. Base to collector and base to emitter response of an *n-p-n* transistor behave like diodes.

As described previously for GFD elements, a DC voltage developed across the load, as charge was unidirectionally pumped onto it (Mantese, Schubring et al. 2001). When a small DC potential was applied at node point A, nearly all the applied DC potential appeared across the GFD because the capacitance of the GFD was substantially smaller than the load. Consequently, an additional charge, q_{GFD}, collected on the GFD in response to the small signal potential; with essentially no charge appeared on the load capacitance due to the applied DC potential. However, a significant amount of charge q_{Load}, collected on the load capacitor because the internal potential of the GFD had been altered. One can, in analogy with Equation 4.1, define a transpacitor beta for the circuit configuration depicted in Figure 4.3 as:

$$q_{Load} = h_{GFD} q_{GFD} \qquad (4.3)$$

Figure 4.5 is a plot of h_{GFD}, as a function of the ratio: DC signal voltage to AC peak supply voltage. Note the authors kept this ratio below 10%. Quite remarkably, h_{GFD}, was between 100 and 200, similar to what is typically found for a transistor device (Horowitz and Hill 1980). Note that charge amplification occurs because the internal potential of the GFD will cause charge to accumulate on the load capacitor until the voltage that develops across the load balances the internal potential of the GFD. Because the load capacitance is many times greater than the GFD, this gain factor can be substantial.

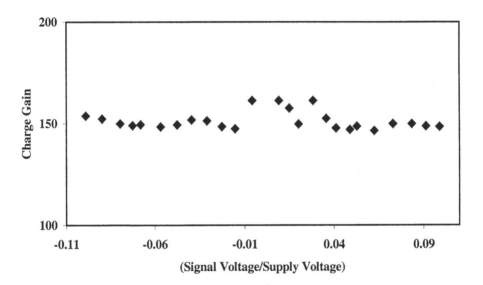

Figure 4.5. h_{GFD} as a function of the ratio: DC signal voltage to AC peak supply voltage (Mantese, Schubring et al. 2001).

Similarly, the additional energy now stored on the load capacitor is:

$$Energy_{Load} = \frac{1}{2C} q_{Load}^2 = \frac{1}{2C} h_{GFD}^2 q_{GFD}^2 \qquad (4.4)$$

which mirrors Equation 4.2.

Of course, a potential applied directly to node A is only one means of altering the internal potential of a GFD. As remarked above, and shown explicitly in Chapter 3, the internal potential is also a strong function of stress and temperature. Thus, it is expected that GFDs in the configuration of Figure 4.3 could be used as a transducer to amplify a small signal thermal or strain input. Figure 4.3 explicitly indicates that other inputs to the internal potential may be realized by external energy sources; we return to this latter point and exploit it in the fabrication of pyroelectric sensors near the end of this chapter. However, in keeping with the discussion of transistor configurations as presented above, we examine the GFD analogue of the *push-pull* transistor amplifier of Figure 4.2.

The same authors who constructed the transpacitive charge amplifier of Figure 4.3 also formed a GFD configuration analogous to the *push-pull* transistor amplifier (Mantese, Schubring et al. 2002). In their construction, they used opposing "up" and "down" devices in separate temperature regulated chambers as shown in Figure 4.6. When the temperatures of the individual devices were adjusted to equalize the outputs of the two GFDs (~27 °C), essentially normal hystereisis was observed. However, if differential heating or cooling was applied to the balanced configuration, a net charge (voltage equal to q_{Load}/C) accumulated on the load capacitor, with a sign dependent on the heating or cooling imbalance.

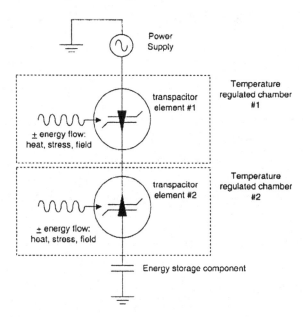

Figure 4.6. Two graded ferroelectric devices in a *push-pull* configuration (Mantese, Schubring et al. 2002).

Figure 4.7 shows that the charge which developed across the load capacitor increased in magnitude with temperature differential. Hence, the transpacitor configuration shown in Figure 4.6 is capable of both nulling out small common mode signals and tracking variations in the local thermal environments of the GFDs. However, while this latter configuration does yield intriguing results, it is still quite simple in design; which suggests the potential for discovering a great many other designs involving both active and passive circuit elements. Finally, it is worthwhile pointing out that these results were independent of the relative placement of the two GFDs. That is, the same response was obtained when the two devices were interchanged but their orientations maintained (Mantese, Schubring et al. 2002). Such is the case, because the observed offsets are the result of internal potentials built into the structures that act to pump charge onto the load capacitor enabling the GFD's to respond to AC charge flow.

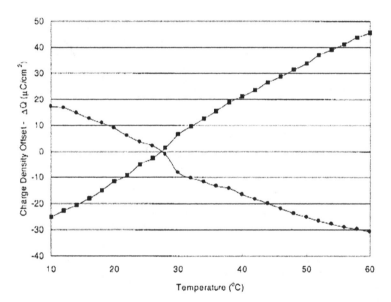

Figure 4.7. Hysteresis displacement (measure in volts across the load capacitor) when: (■) the temperature of transpacitor No. 2 was varied and transpacitor No. 1 was held fixed at 27°C, and (●) when the temperature of transpacitor No. 1 was varied and transpacitor No. 2 was held fixed at 27°C (Mantese, Schubring et al. 2002).

4.   PYROELECTRIC SENSORS

The spontaneous polarization, P_s, of a ferroelectric can be a very strong function of temperature near its Curie point, T_c, where a ferroelectric material transitions from a centro-symmetric structure to one that supports a spontaneous polarization. The typical temperature dependence of the polarization is illustrated in Figure 4.8. This abrupt characteristic is primarily observed for bulk ceramic materials, and to a lesser extent for thin film ferroelectrics. Consequently, when a GFD is constructed from compositions with a mean temperature T_c, as shown in Figure 4.9, the internal potential of the GFD becomes likewise highly temperature dependent around the average T_c. The temperature dependence of the internal potential, together with the charge gain characteristics of transpacitors can be exploited to good effect in the construction of infrared sensitive devices, i.e., pyroelectric detectors.

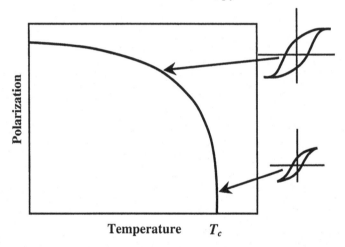

Figure 4.8. Plot showing the typical variation of a ferroelectric material's spontaneous polarization with temperature near its Curie point.

Thermal detectors formed from homogeneous composition ferroelectric materials, have a long and interesting history. A number of excellent reviews of infrared (IR) ferroelectric pyrometers are available in the literature, with the most sensitive devices constructed from bulk BST and integrated with silicon via bump technologies (Whatmore 1986; Whatmore, Osbond et al. 1987; Kulwicki, Amin et al. 1992). A great deal of the discussion concerning ferroelectric thermal detectors has focused on the pyroelectric coefficient, p, of the underlying material; which is defined as:

$$p = dP_s / dT + \varepsilon_o \int_0^E (\partial \varepsilon / \partial T)_E \, dE \qquad (4.5)$$

where P_s is the spontaneous polarization, T is the temperature, ε is the field-dependent permittivity, and E is the applied electric field. p, thus represents the change in accumulated charge per unit area, with unit temperature change. For optimum sensitivity and performance, one would therefore choose materials and circuit configuration that take small energy inputs in the form of temperature changes and converts them into large charge signals.

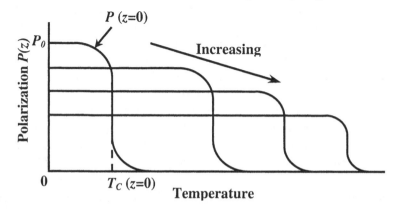

Figure 4.9. Idealization of the temperature dependence of the polarization at various positions for a ferroelectric material with a polarization gradient normal to the growth surface.

The conventional pyroelectric coefficient contains two parts: (1) That due to the change in spontaneous polarization, when the detector is operated below the Curie temperature (the first term in the above equation). (2) And a second part (the pyroelectric bolometric term), which is the field-induced change in charge on a capacitor with temperature change when the device is operated at slightly above the Curie temperature. It should be noted that the two terms are of opposite sign.

For BST, the term due to the change in spontaneous polarization is often less than 1 $\mu C/cm^2$-°C even for bulk single crystal materials, with most device work reported from bulk ceramics yielding values $\leq$ 0.5 $\mu C/cm^2$-°C (Whatmore 1986; Whatmore, Osbond et al. 1987; Kulwicki, Amin et al. 1992). The pyroelectric coefficient for the bolometric term, however, can reach exceedingly large values; as great as 23 $\mu C/cm^2$-°C (Kulwicki, Amin et al. 1992), though lesser values are often observed. These latter values are deceiving, for when the devices are operated in the bolometric mode at temperatures slightly above the Curie temperature, ferroelectrics become lossy because tan δ becomes large, which introduces significant noise in the detector elements (Whatmore 1986). Thus, substantial bias field must be applied to the dielectric to suppress tan δ, otherwise losses and noise are too great. When applying this bias, the bolometric pyroelectric coefficient drops rapidly because ε (the temperature dependent permittivity) is a non-linear function of applied field (Whatmore 1986). Indeed, most commercial devices are constructed from bulk BST and operate with p less than 1 $\mu C/cm^2$-°C.

It would be wrong, however, to conclude from the above discussion that the pyroelectric coefficient is the only parameter of interest for thermal detectors. Such is not the case, as there are a great many figures of merit the detector community uses to characterize both the materials and systems. Indeed, system issues often dominate the cost and performance of IR detectors. However, the pyroelectric coefficient is often focused on, as a large value is usually a necessary condition for good device performance.

While IR detectors formed from bulk BST ceramics yield impressive imagery with noise equivalent temperatures (NEΔT's) of 0.05 °C or less, they are costly to produce because they are fabricated using a hybrid process, marrying bulk ceramic and IC processing. An obvious choice would therefore be to integrate thin film ferroelectrics directly into the IC process. Unfortunately, however, thin film ferroelectric materials have greatly reduced pyroelectric properties, usually on the order of 0.001-0.05 $\mu C/cm^2$-°C, due to built in strains when integrated with conventional IC materials; making them unsuitable for conventional pyroelectric sensors.

We, therefore, again turn to the construct of Figure 4.3, which we demonstrated was capable of charge gain, as captured in Equation 4.3. We now imagine the input to the charge amplifier to be an external IR energy source that alters the internal potential of the GFD. Once again, hysteresis offsets were observed, which were seen to be strong functions of GFD temperature. Infrared radiation, whose effect was to vary the temperature of the GFD was simulated using a temperature-controlled chamber capable of 0.05 °C regulation. Figure 4.10 shows the charge accumulated across a load capacitor as a function of temperature for a KTN GFD configured as in Figure 4.3, using a "down" GFD (Schubring, Mantese et al. 1992; Jin, Auner et al. 1998; Mohammed, Auner et al. 1998). Note the strong temperature dependence near room temperature. From Figure 4.10 an effective pyroelectric coefficient may be constructed as:

$$p_{eff} = (1/A)(\Delta Q / \Delta T) \qquad (4.6)$$

where A is the area of the GFD and $(\Delta Q/\Delta T)$ is the change in DC charge offset that appeared on the load capacitor with temperature. p_{eff} is plotted in Figure 4.11. Note that at its maximum, p_{eff} exceeds 10 $\mu C/cm^2$-°C, superior to even the very best single crystal and bulk ceramic pyroelectric materials, as this is true charge accumulation on the load capacitor (Schubring, Mantese et al. 1992). Similar results have been obtained for polarization graded BST thin film ferroelectrics(Jin, Auner et al. 1998; Mohammed, Auner et al. 1998).

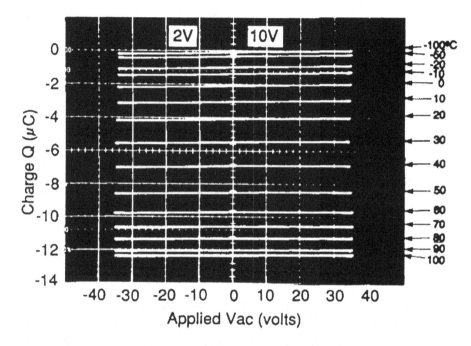

Figure 4.10. Variation in charge offset as a function of temperature for a KTN transpacitive element (Schubring, Mantese et al. 1992). Reprinted with permission from [Schubring, N. W., Mantese, J. V., Micheli, A. L., Catalan, A. B. and Lopez, R. J., "Charge Pumping and Pseudopyroelectric Effect in Active Ferroelectric Relaxor-Type Films," *Physical Review Letters* **68**, 1778-1781, (1992)]. Copyright 1992 by the American Physical Society.

To date, no pyroelectric focal plane arrays (FPA's) have been fabricated from polarization-graded ferroelectrics. But the above findings do illustrate that the concepts outlined in this chapter are consistent with the theoretical analysis of this book and are an indication of the general utility of the construct depicted in Figure 4.3.

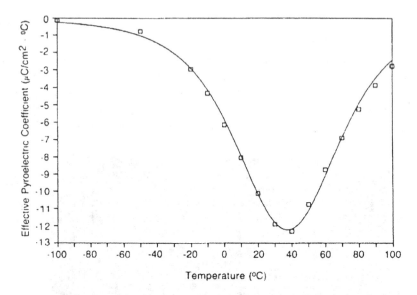

Figure 4.11. Effective pyroelectric coefficient (Schubring, Mantese et al. 1992). Reprinted with permission from [Schubring, N. W., Mantese, J. V., Micheli, A. L., Catalan, A. B. and Lopez, R. J., "Charge Pumping and Pseudopyroelectric Effect in Active Ferroelectric Relaxor-Type Films," *Physical Review Letters* **68**, 1778-1781, (1992)]. Copyright 1992 by the American Physical Society.

5.   OTHER TRANS-CAPACITIVE DEVICES

Polarization graded ferroelectrics have only been used in the assemblage of the most simple trans-capacitive circuits and pyroelectric sensors (Schubring, Mantese et al. 1992; Schubring, Mantese et al. 1999; Mantese, Schubring et al. 2001; Mantese, Schubring et al. 2002). Thus, there are a great many unexplored configurations consisting of only GFD elements and constructs where transpacitive devices are creatively integrated with transistors and other more advanced IC. These latter configurations can be made even more simply when GFDs can be routinely fabricated with reproducible device performance parameters. Undoubtedly, any new trans-capacitive devices will also exploit other couplings to the internal potentials of GFDs, making this an exciting area for future research and discovery.

Chapter 5

GRADED FERROICS AND TRANSPONENTS

1.   INTRODUCTION

Can the concepts developed in the previous chapters of this book be generalized to other ferroic systems? If so, what are the essential aspects of graded ferroics and what can be expected from their corresponding active devices, "transponents"? These are the questions we aim to address in this final chapter. Essentially, we will expand the idea of grading to other functional systems. Central to all the ferroics we will consider are those systems that can be characterized by a Landau-Ginzburg potential with a fixed gradient in order parameter and which display hysteresis. Both elements lead to logical extensions of graded ferrolectrics and transpacitors. However, at the end of chapter 5 we further extend our analysis to the general class of "smart" materials.

Ferroics form an essential sub-group of functional materials whose physical properties are sensitive to changes in external conditions such as temperature, pressure, electric, and magnetic fields (Aizu 1970; Wadhawan 2000). Ferroelectric, ferromagnetic, and ferroelastic materials are the best known examples of ferroics that are principally distinguished by four main characteristics (Wadhawan 2000). First, their property-specific order parameter(s) (e.g., polarization, magnetization, or self-strain, for ferroelectrics, ferromagnets, and ferroelastics, respectively) spontaneously assume non-zero values below a threshold temperature even in the absence of an applied stimulus. The emergence of such an order parameter may be due to a displacive transformation, (electrical or magnetic) ordering, or a combination of both. This transition is accompanied by a structural phase change from a centro-symmetric crystalline structure, to one in which there is an asymmetry (spontaneous symmetry breaking). Such behavior is a direct result of the nature of the underlying internal potentials, which are characterized by double well minima as shown in Figure 5.1. Here, two possible response states are shown for the three most common ferroic systems. Secondly, ferroics exhibit hysteresis in their stimulus-response behavior: e.g., polarization vs. applied electric field, magnetization vs. applied magnetic field, and strain vs. applied stress. Figure 5.2a shows the response of these latter three material systems with a generalized hysteresis graph for any ferroic system. The hysteretic characteristics of the stimulus-response behavior of ferroic systems are due to domain phenomena. Another distinguishing characteristic of ferroic materials is their large and non-linear generalized susceptibilities. For example, the dielectric response, the magnetic susceptibility, and the elastic moduli in ferroelectric, ferromagnetic, and ferroelastic materials, respectively, show an λ-type critical behavior near the phase transformation temperature T_C (Figure 5.2b). And finally, the ferroic phase transformation can be induced via an external field conjugate to the order parameter. This field in ferroelectrics and ferromagnetics are an applied electric (**E**) and magnetic (**H**) field, respectively. In ferroelastics, a super-elastic deformation can be induced by an applied stress in the material in its austenitic (paraelastic) phase. The special interphase boundaries between the austenite and the stress-induced martensitic phases provide an easy mode of deformation via reversible interphase and domain wall motion which may result in an elastic deformation close to 10% in certain Cu-Al-Ni and Ni-Ti alloys (Nishiyama 1978; Otsuka 1998).

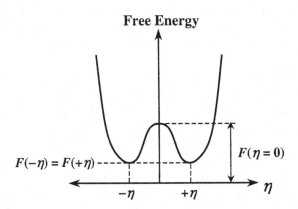

Figure 5.1. Double-well potential for a typical ferroic material below T_C showing equilibrium states characterized by $dF/d\eta = 0$. The order parameter η corresponds to polarization, magnetization, and self-strain of a ferroelectric, ferromagnetic, and ferroelastic material, respectively.

Ferroics substances are usually high energy-density materials that can be configured to store and release energy (electrical, magnetic, and mechanical) in a well controlled manner, making them highly useful as sensors and actuators. The term ferroic has also been extended to include ferrogyrotropic materials, and other unconventional systems such as high-T_C and BCS superconductors, superfluids, microemulsions, materials near critical points, and even giant and colossal magnetoresitive (GMR and CMR) materials

(Wadhawan 2000). Other examples may be constructed under the general definitions given above.

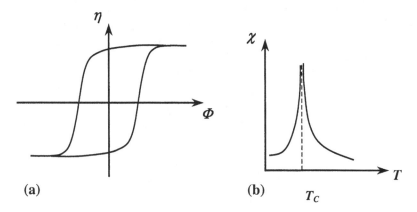

Figure 5.2. (a) Stimulus-response behavior of a typical ferroic. Φ refers to the external field conjugate to the order parameter, η. (b) Ferroic susceptibility versus temperature curve displaying anomaly near the phase transformation temperature. χ corresponds to dielectric susceptibility, magnetic susceptibility, and elastic compliance in a ferroelectric, ferromagnetic, and ferroelastic material, respectively.

While extensively studied both theoretically and experimentally, ferroic research has by-and-large been confined primarily to investigating the properties of homogeneous and layered sub-systems. Considering the similarity of many ferroic systems, we attempt to develop a common understanding for graded ferroic devices and structures. Our analysis of Chapter 2 pertaining to polarization-graded ferroelectrics is therefore expanded qualitatively in this chapter to a discussion centered about a generalized thermodynamic theory which can be used to develop a methodology for analyzing graded ferroics. Material system inhomogeneities are assumed to arise from compositional, temperature, or stress gradients. As we will show, these spatial non-uniformities result in local order parameters having corresponding spatial variation. Based on the discussion in Chapter 2, we expect that ferroics, in general, to possess non-uniform free energies with attendant internal potentials; the latter of which should be reflected by displacements of the material's stimulus-response hysteresis plots along the response axis (e.g., polarization, magnetization, or strain axis).

2.   GENERALIZED THEORY

Let us consider the expansion of the free energy of a ferroic phase transformation of a single-domain system with three order parameters η_i such that it is a harmonic function of the order parameters (and thus does not contain odd powers):

$$G(\eta_i) = \int_V \left[\alpha_{ij}\eta_i\eta_j + \beta_{ijkl}\eta_i\eta_j\eta_k\eta_l + A_{ijkl}\left(\nabla_i\eta_j \cdot \nabla_k\eta_l\right) + \ldots \right.$$

$$+ \delta_{ijk}x_{ij}\eta_k + \frac{1}{2}q_{ijkl}x_{ij}\eta_k\eta_l \qquad (5.1)$$

$$\left. -\left(\frac{1}{2}\Phi_i^D - \Phi_i\right)\eta_i + F_{el,T}\left(C_{ijkl}\right)\right]dV$$

where α_{ij}, β_{ijkl}, and A_{ijkl} are the free energy expansion coefficients, δ_{ijk} and q_{ijkl} are the bilinear and linear-quadratic coupling coefficients between the order parameter and the strain x_{ij}, C_{ijkl} are the elastic coefficients, Φ_i is an externally applied electrical or magnetic field, and Φ_i^D is the internal depolarization or demagnetization field for ferroelectric of ferromagnetic materials systems, respectively. In Equation 5.1, $F_{el,T}$ is the total elastic energy which has two components:

$$F_{el,T}\left(C_{ijkl}\right) = F_{el} + \frac{1}{2}\sigma_{ij}x_{kl} \qquad (5.2)$$

where F_{el} is the internal strain energy due to variations in the self strain as discussed in Chapter 2 and the last term is the elastic energy of an applied stress field, σ_{ij}. As will be shown in this chapter, proper ferroelectric, ferromagnetic, and ferroelastic phase transformations can be described via the above relation with the polarization P_i, magnetization M_i, or the self-strain e as the order parameter, respectively.

We will limit ourselves for the time being to graded ferroelectric or ferromagnetic systems. Graded ferroelastics where the self-strain itself is the order parameter will be treated in subsequent sections. Consider a graded ferroelectric or ferromagnetic bar of length L with a singular order parameter η and a systematic spatial variation along L such that $\partial\eta/\partial z \neq 0$. Here, we make the gradient term large but constant over the spatial dimensions of our system, which allows us to solve the above relation generally; unlike in systems near their critical point, where large fluctuations in order parameter force one to keep higher order $\partial \eta / \partial z$ expansion terms, making the problem intractable. It is precisely this gradient term which provides the basis for the hysteresis offsets with respect to the conjugate field, Φ, of the order parameter.

$$G(\eta,T) = \int_0^L \left[\frac{1}{2}\alpha(T-T_C)\eta^2 + \frac{1}{4}\beta\eta^4 + ... + \frac{1}{2}A\left(\frac{\partial \eta}{\partial z}\right)^2 \right.$$
$$\left. -\left(\frac{1}{2}\Phi_i^D - \Phi_i\right)\eta + F_{el,T}(C_{ijkl})\right] dz. \qquad (5.3)$$

For the sake of simplicity, we assume that η is only a function of position and varies linearly along z:

$$\frac{\partial \eta}{\partial z} = \frac{\eta(z=L) - \eta(z=0)}{L} \equiv \frac{\Delta \eta}{L} = const.. \qquad (5.4)$$

Further, it is assumed that there are no externally applied fields, i.e., $\Phi=0$ and $F_{el,T}=F_{el}$. A conjugate field may be defined as:

$$\frac{\partial G}{\partial z} = \Phi = \alpha(T-T_C)\eta + \beta\eta^3 - A\frac{d^2\eta}{dz^2} - \frac{1}{2}\Phi_i^D + \frac{dF_{el}}{d\eta}. \qquad (5.5)$$

The gradient in the order parameter and the resulting depoling field as well as the elastic energy result in a symmetry breaking that leads to skewed potential wells much like Figure 2.21 (see Figure 5.3). From the definition of this "built-in" conjugate field we have:

$$\int_0^L \Phi(z) \cdot dz = \int_0^L \frac{\partial G(z)}{\partial \eta(z)} \cdot dz = \left(\frac{d\eta}{dz}\right)^{-1} \int_0^L \frac{\partial G(z)}{\partial \eta(z)} \frac{d\eta}{dz} \cdot dz , \qquad (5.6)$$

or,

$$\int_0^L \Phi(z) \cdot dz = \left(\frac{d\eta}{dz}\right)^{-1} [G(\eta(z=L)) - G(\eta(z=0))] = \frac{L}{\Delta\eta}\Delta G , \qquad (5.7)$$

from which we conclude the existence of skewed potential in the Landau-Ginzburg representation of a graded ferroic; thereby yielding a preferred value for the two-state order parameter, Figure 5.3. Consequently, an asymmetry reveals itself as an offset in an otherwise normal hysteresis curve, Figure 5.1. We note that the case of graded ferroelastics, the built-in field is due to a position dependent self-strain that results in bending.

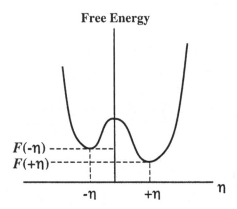

Figure 5.3. Skewed potential in the presence of an internal built-in field due to a systematic gradient in the order parameter η.

In analogy with graded ferroelectrics, one can describe generalized "transponent" devices similar to the transpacitive structures of Chapter 4. Here, however, the transponents are excited by their conjugate fields with variations in internal potential affected by external variations such as temperature, magnetic intensity, angular velocity, or stress. Amplification requires an additional "linear" passive element capable of storing $\frac{1}{2}\Phi\eta$ energy.

In the sections that follow, we discuss a number of ferroic systems. The selection is by no means exhaustive. In Table 5.1 above we compare the general features of some graded ferroics. It is obvious from this presentation, however, that the concepts developed here may be extended to other ferroics not covered in the discussions which follows.

Table 5.1. Comparison of the general features of graded ferroics

System	Order Parameter	Conjugate Field	Gradient Parameters	Transponent Device	Potential Application
Ferroelectric	$P(r)$, Electric Polarization	E, Electric Field	T, c, σ	Transpacitor	IR Sensor
Ferromagnetic	$M(r)$, Magnetic Polarization	H, Magnetic Intensity	T, c, σ	Transductor	Magneto-meter
Ferroelastic	e, Strain	σ, Stress	T, c	Translastic	Strain sensor
Superconductor	$\Psi(r)$, Pairing Function	Pair Source	T, H	?	?

3. FERROMAGNETS

Ferromagnetism is a phenomenon by which a material can exhibit a spontaneous magnetization, and is one of the strongest forms of magnetism (Bozoroth 1978; Jiles 1998). It is responsible for most of the magnetic behavior encountered in everyday life, and is the basis for all permanent magnets (as well as the materials that are attracted to them).

Ferromagnetism can be understood by the direct influence of two effects from quantum mechanics; spin of an electron around its own axis and the Pauli exclusion principle (Hummel 2001). The spin of an electron has a magnetic dipole moment and creates a magnetic field. In many materials (specifically those with a filled electron shell), however, the electrons come in pairs of opposite spin, which cancel one another's dipole moments. Only atoms with unpaired electrons (partially filled shells) can experience a net magnetic moment from spin. A ferromagnetic material has many such electrons, and if aligned they create a measurable macroscopic field.

The spins/dipoles tend to align parallel to an external magnetic field, an effect called *paramagnetism*. In ferromagnetic materials, below a critical temperature T_C, the spins tend to align spontaneously, without the application of an external field. This magnetic ordering is a quantum mechanical effect. According to classical electromagnetism, two nearby magnetic dipoles will tend to align in opposite directions, which would create an *antiferromagnetic* material. In a ferromagnet, however, they tend to align in the same direction because of the Pauli principle: two electrons with the same spin cannot lie at the same position, and thus feel an effective additional repulsion that lowers their electrostatic energy. This difference in energy is called the exchange energy given by:

$$F_{Ex} = -2J_{Ex} \mathbf{S}_i \cdot \mathbf{S}_j \qquad (5.8)$$

where J_{Ex} is the exchange integral and $\mathbf{S}_i$ and $\mathbf{S}_j$ are the spins on neighboring atoms i and j, respectively. For a ferromagnetic material, J_{Ex} is positive and induces nearby electrons to align along the same direction. If J_{Ex} is negative an antiparallel alignment is preferred resulting in antiferromagnetism. Thermal agitations counteract this magnetic ordering. The interplay between the energy of atomic vibrations and the exchange energy results in the destruction of the spontaneous alignment of magnetic dipoles and gives rise to a disordered state with no net magnetization. These dipoles, however, can still be aligned but only in the presence of an applied magnetic field and thus ferromagnetic materials are paramagnetic above the transition temperature.

One fundamental issue in ferromagnetism is related to domain phenomena (Hubert 1974; Hubert and Schafer 1998). Consider a uniformly magnetized ferromagnetic crystal with all the spins pointing in one direction as shown in Figure 5.4a. This configuration would result in the creation of north and south poles at the opposite ends of the crystal. This obviously results in a large magnetic field opposing the direction of magnetization in the crystal. This demagnetizing field can be reduced by the formation of magnetic domains as shown in Figure 5.4b. The domain configuration in Figure 5.4b cannot completely eliminate the demagnetization although the magnetic field is now confined to a smaller volume. The magnetostatic energy of the depolarization can further be reduced by the formation of additional magnetic domains resulting in a completely closed path within the crystal as to eliminate the occurrence of magnetic poles (Figure 5.4c).

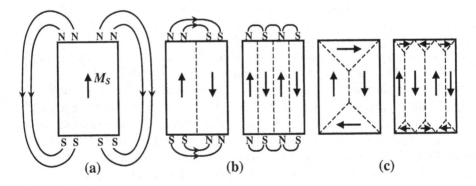

Figure 5.4. The formation of magnetic domains to minimize the demagnetization energy in a uniformly magnetized ferromagnetic crystal (Jiles 1998). Reprinted with permission from [Jiles, D., *Introduction to Magnetism and Magnetic Materials*, London, Chapman and Hall (1998)]. Copyright CRC Press, Boca Raton, Florida.

The transition between two domains, where the magnetization flips, is called a Bloch wall (Figure 5.5), and is a gradual transition on the atomic scale (covering a distance of about 100 nm for iron) (Kittel 1996). The energy and the width of the wall are determined by the strength of the exchange energy and the anisotropy energy which points the magnetization along certain crystallographic axes called directions of easy magnetization (Kittel 1946). In its simplest form the anisotropy energy for a cubic symmetry can be expressed approximately as:

$$F_A = K_1(\cos^2 \theta_1 \cos^2 \theta_2 + \cos^2 \theta_2 \cos^2 \theta_3 + \cos^2 \theta_3 \cos^2 \theta_1) \qquad (5.9)$$

where K_1 is the first-order anisotropy constant for a cubic system and $\cos\theta_i$ are the direction cosines of the magnetization with respect to the crystal axes.

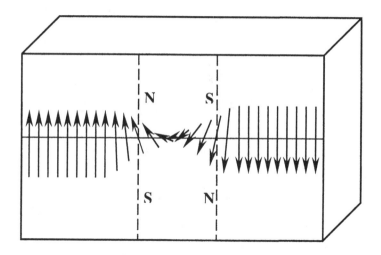

Figure 5.5. Alignment of the magnetization in a Bloch wall (Jiles 1998). Reprinted with permission from [Jiles, D., *Introduction to Magnetism and Magnetic Materials*, London, Chapman and Hall (1998)]. Copyright CRC Press, Boca Raton, Florida.

The magnetic response of a ferromagnetic material is related to domain wall motion and rotation of the magnetization direction. The magnetization **M** (the magnetic dipole density) as a function of an applied field **H** results in a hysteresis that looks much alike to the ones for ferroelectrics and ferroelastics (Figure 5.6). The "virgin" state has no net polarization although it consists of domains, which are uniformly magnetized. The magnetization increases with an applied field until saturation is reached corresponding to complete switching of domains. If the field is increased, the magnetization may increase further (induced magnetization) but this increase is usually small. When the applied field is reversed, domains oriented in an opposite direction with respect to the applied field nucleate and grow via domain wall motion. If the field is applied in a direction not along the magnetization direction, the initial stage of magnetization (low field) is through reversible domain wall displacements and the final stage (high field) involves the rotation of the magnetization within the crystal to a direction parallel to the applied field. A complete treatment of domain phenomena in ferromagnets can be found in Hubert's classical analysis, *Theorie der Domänenwände in Geordneten Medien* (Hubert 1974) and *Magnetic Domains: The Analysis of Magnetic Microstructures* (Hubert and Schafer 1998).

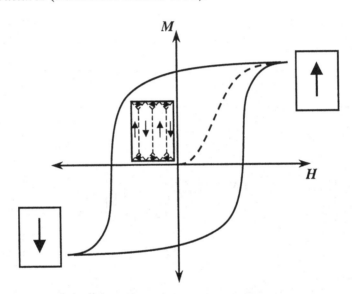

Figure 5.6. A typical magnetization hysteresis curve of a ferromagnetic material and the schematic domain arrangements at critical fields.

Associated with magnetization of a crystal, there are changes in the crystal dimensions. This effect is called magnetostriction, and it is one of the important magnetic properties which accompany ferromagnetism (Jiles 1998). In formal treatments, a magnetostrictive coefficient λ is defined as the fractional change in length l as the magnetization increases from zero to its saturation value, dl/l. The magnetostriction defines an elastic strain associated with the deformation due to alignment of the magnetic dipoles as well as dimensional changes due to induced magnetization if the dipoles are completely aligned. This property, which allows magnetostrictive materials to convert magnetic energy into mechanical energy and *vice versa*, is used for the actuation and sensing applications. The saturation magnetostriction in a cubic crystal via the magnetostrictions along crystal axes are given by (Jiles 1998):

$$
\begin{aligned}
\lambda_S = {} & \frac{3}{2}\lambda_{100}(\alpha_1^2\beta_1^2 + \alpha_2^2\beta_2^2 + \alpha_3^2\beta_3^2 - \frac{1}{3}) \\
& + 3\lambda_{111}(\alpha_1\alpha_2\beta_1\beta_2 + \alpha_2\alpha_3\beta_2\beta_3 + \alpha_3\alpha_1\beta_3\beta_1)
\end{aligned}
\tag{5.10}
$$

where λ_{110} and λ_{111} are the magnetostrictions (that quantify the self-strain due to magnetization) along <110> and <111> directions, respectively, and α_i and β_i are the direction cosines relative to the applied field direction and the direction along which the magnetic moments are saturated and the direction along which the magnetization is measured, respectively. The self-strains along the <100> and <111> directions are thus given by $e_{111}=(3/2)\lambda_{100}$ and $e_{111}=(3/2)\lambda_{111}$, respectively.

In the majority of oxides, the exchange energy is negative leading to anti-parallel alignment between spins of neighboring atoms. If the strength of the magnetic moments in these sublattices is equal, the material is said to be antiferromagnetic. Ferrimagnetism results when the magnetizations in the sublattices are not equal. These materials behave in the macroscale like ferromagnets. They possess a spontaneous polarization and a domain structure below a critical temperature. Similar to ferromagnets, magnetic switching is hysteretic. But unlike metallic ferromagnets, they are electrical insulators, making these materials very attractive for high-frequency applications where a large resistivity is required (Smit and Wijn 1959). The single-crystal spontaneous magnetizations, the transition temperatures, and the magnetostrictions of typical ferromagnetic materials are summarized in Table 5.2.

Thermodynamic models based on the Landau approach can be used to describe the ferromagnetic phase transformation. In its simplest form, the Landau potential of a second order ferromagnetic phase transformation is given by (Kadanoff, Gotze et al. 1967; Harrison 2000):

$$
\begin{aligned}
G(M(\mathbf{r}),T) = {} & G_0 + \frac{1}{2}A(T - T_C)M(\mathbf{r})^2 + \frac{1}{4}BM(\mathbf{r})^4 \\
& + \frac{1}{2}D[\nabla M(\mathbf{r}) \cdot \nabla M(\mathbf{r})]
\end{aligned}
\tag{5.11}
$$

where $\mathbf{M}=M(\mathbf{r})=[m_x, m_y, m_z]$ is the vector of magnetization, T_C is the Curie temperature, and A, B, and D are the expansion coefficients. $M(\mathbf{r})$ describes magnetic ordering such that $|\mathbf{M}=M(\mathbf{r})|/M_S=1$ for the completely ordered state and $|\mathbf{M}=M(\mathbf{r})|/M_S=0$ for the paramagnetic phase above T_C, where M_S is the saturation magnetization. The last term in the equation serves to damp out spatial variations in $\mathbf{M}$ and becomes significant near the phase transformation temperature (Kadanoff, Gotze et al. 1967).

Table 5.2. The single-crystal saturation (spontaneous) magnetizations, the Curie temperatures, and the saturation magnetostrictions of typical ferromagnetic materials.

Substance	Saturation Magnetization M_s, in gauss (at room T)	Curie Temperature, in K	Saturation Magnetostriction (10^{-6})
Terfenol-D	10000	653	1020
Fe	1707	1043	-4.2
Ni	485	627	-32.8
Co	1400	1388	-14.4
$MnFe_2O_4$	400	573	-5
Fe_3O_4	480	858	40
$CoFe_2O_4$	425	793	-110
$NiFe_2O_4$	270	858	-26
$MgFe_2O_4$	120	713	-6
$Li_{0.5}Fe_{2.5}O_4$	310	943	-8

As is the case with any ordering phenomena, the Landau model outlined above is a mesoscopic description and thus is insensitive to short-range ordering. It ignores the fact that two configurations with the same global order parameter may have different short-range ordering. Considering that exchange energy is mostly determined by nearest-neighbor interactions, such a mean-field approximation is often inadequate to describe the fundamental physical mechanisms of ferromagnetism. Albeit this shortcoming, the Landau model is simple and is easily manipulated. For a more rigorous understanding, two- and three-dimensional Ising models or quantum mechanical approaches may be employed.

For the sake of simplicity, we will use the Landau expansion of the free energy to describe both graded and non-graded ferromagnets. Consider a compositionally graded single-crystal ferromagnetic bar of length L. The grading is along the length L (z-direction) which coincides with the easy magnetization direction of a ferromagnet with $\mathbf{M}=[M,0,0]$. In such a case, very similar to graded ferroelectrics and ferroelastics, the Landau potential takes the form:

$$G = \int_0^L \left[\frac{1}{2} A(z)[T - T_C(z)]M(z)^2 + \frac{1}{4} B(z)M(z)^4 + \frac{1}{2} D\left(\frac{dM(z)}{dz}\right)^2 \right.$$

$$\left. - \frac{1}{2} N_D M(z)^2 + F_{el}[C_{ij}(z), \lambda_{ij}(z), M(z)] \right] dz \tag{5.12}$$

where $(1/2) \cdot N_D \cdot M(z)^2 = H_D \cdot M(z)$ is the magnetostatic energy due to the demagnetization, H_D is the demagnetizing field, and N_D is a demagnetization factor that depends only on the geometry of the sample. The elastic energy due to the magnetization gradient is a function of the elastic constants C_{ij}, the magnetostrictions λ_{ij}, and the magnetization. The magnetization gradient, similar to graded ferroelectrics, translates into a built-in magnetic field. According to Gauss law:

$$\nabla \cdot \mathbf{B} = 0 \tag{5.13}$$

where $\mathbf{B}$ is the vector of magnetic inductance given by:

$$\mathbf{B} = \mu_0(\mathbf{H} + \mathbf{M}). \tag{5.14}$$

Therefore a magnetization gradient should result in a built-in magnetic field:

$$\nabla \cdot \mathbf{H} = -\nabla \cdot \mathbf{M}. \tag{5.15}$$

We note that F_{el} can be obtained through the conditions that the average internal stress and the average momentum of the internal stress should be zero. Again, similar to ferroelectrics, such an effect can be expected were the grading achieved through temperature or stress gradients since $M = f(c, T, \sigma_{ij})$.

Unlike ferroelectrics, there is no compensation mechanism in ferromagnets for the demagnetization. Therefore, the graded ferromagnet will undoubtedly split spontaneously into magnetic domains. To harness the built-in field due to the magnetization gradient, the ferromagnet has to be magnetized externally to overcome the demagnetizing field H_D that in essence corresponds to magnetizing the material above the coercive field of the "hardest" ferromagnetic layer.

Although non-homogeneous (graded) ferromagnetics have received some attention in the past, the majority of those studies concentrated on achieving a spread in Curie temperatures across the sample as a means of smoothing the abrupt change in susceptibility that occurs near a phase transition, see Figure 5.2b (Wagner, Romanov et al. 1996). Similar approaches have also been used for ferroelectrics as a means of making capacitive elements less temperature sensitive. However, in one study, the hysteresis loops of compositionally graded 30 nm TbFe films (obtained using both Kerr and Hall effect measurements) have shown an anomalous behavior compared to a double layer TbFe film of the same thickness (Chen, Malmhall et al. 1983). The authors have attributed this to a built-in "...*magnetostatic energy gradient to drive the compensation wall through the film.*" Unfortunately, this work was not expanded upon or pursued further.

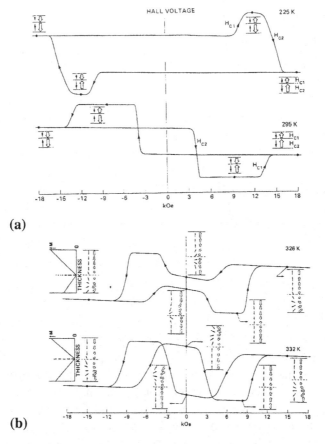

(a)

(b)

Figure 5.7. (a) Anomalous hysteresis loops in a double layer TbFe film measured by Hall effect at 225 and 295 K. The open arrows indicate the direction and magnitude of the net magnetization of each layer, whereas the solid arrow shows the direction of the Tb sublattice moment. (b) Schematics of the net magnetization (open arrows) and Tb sublattice moment (solid arrows) for different state of magnetization in a compositionally graded TbFe film at 326 and 332 K (Chen, Malmhall et al. 1983) Reprinted from "Anomalous Hysteresis Loops in Single and Double Layer Sputtered TbFe Films" *Journal of Magnetism and Magnetic Materials*, Chen et al., 35, 517-543, Copyright 1983, with permission from Elsevier.

The operation and applications, therefore, of active graded ferromagnetics, "transductors", is unknown as of this writing as no such devices have been constructed or explored theoretically. However, in analogy with transpacitor elements (which have been extensively studied experimentally) and by referring to Figures 1.9 and 4.3; it is obvious that transponent devices must be comprised of additional elements other than a magnetization graded ferromagnetic element; including: (1) A source of external signal, e.g., temperature, strain, magnetic flux. (2) An external driving potential, i.e., an energy source or reservoir; in this case an applied periodic magnetic field. (3) A linear energy storage element, for example an inductor or possibly a capacitor.

While entirely speculative, in analogy with our transpacitor *emitter-follower-like* amplifier, it is conjectured that transductors may be able to operate as high sensitivity: magnetometers, strain gauges, and magneto-caloric or magneto-electric devices. It may also be true that because there is not the analogue of free charge in magnetic circuits, one can also operate such devices in a dc mode; but this too is speculation at this point. Clearly, much is unknown about compositionally graded ferromagnetic systems. Hardly any literature refers to such systems experimentally or theoretically. Hence such systems offer rich areas of fundamental research with unknown potential applications and technical impact.

4. FERROELASTICS

What are ferroelastics? Following Salje's definitive book on ferroelastic crystals (Salje 1990), we find, not surprisingly, a description that is quite similar to the general definition of ferroic materials. According to Salje, there are two necessary ingredients for a material to be ferroelastic: "*...a phase transition between the paraelastic and the ferroelastic phase which creates a lattice distortion*" and "*...this lattice distortion can be reoriented by external stress.*" The lattice distortion or the spontaneous (self) strain is the order parameter of ferroelastic phase transformations. The reorientation of the lattice distortion (or ferroelastic "switching") is accomplished via reversible domain wall movements in the presence of an applied stress.

Similar to ferroelectrics and ferromagnets, the formation of domains is due to the minimization of an internal field. As we have discussed in Chapter 2, the depoling field due to uncompensated charges will result in the formation of electrical (or 180°) domains in ferroelectrics. Likewise, the formation of transversely modulated structures in epitaxial fields has been predicted on the basis of the concept of elastic domains, i.e., structural domains which reduce the intensity of the elastic field and elastic energy in a manner similar to that in which magnetic (or electric) domains in ferromagnetics (or ferroelectrics) reduce magnetic (or electric) fields.

The theory of elastic domains has been developed and successfully applied to describe different modulated structures in bulk materials and thin films (Roitburd 1976; Roytburd 1993; Roytburd 1998). The formation of domains (or twins) in ferroelastic materials minimizes internal elastic stresses that may exist in materials due to a number of reasons. The twins may form during nucleation and growth or may be induced via external stresses. Twinning, or polydomain formation, may occur in several technologically important materials systems including high-T_C superconductors, several perovskite oxides, and a large number of metallic alloy systems.

Consider a cubic to tetragonal ferroelastic phase transformation. The self-strains may be expressed similar to Equation 2.5 as:

$$\varepsilon_1^0 = \begin{pmatrix} e & 0 & 0 \\ 0 & e' & 0 \\ 0 & 0 & e' \end{pmatrix}, \varepsilon_2^0 = \begin{pmatrix} e' & 0 & 0 \\ 0 & e & 0 \\ 0 & 0 & e' \end{pmatrix}, \varepsilon_3^0 = \begin{pmatrix} e' & 0 & 0 \\ 0 & e' & 0 \\ 0 & 0 & e \end{pmatrix}, \qquad (5.16)$$

where $e=(c-a_0)/a_0$, $e'=(a-a_0)/a_0$, a and c are the lattice parameters in the ferroelastic state and a_0 is the lattice parameter in the paraelastic state. These three orientational variants of the ferroelastics state are shown in the inner circle of Figure 5.8 (Roytburd, Alpay et al. 2001; Alpay 2002). Depending on the strength of internal stresses and constraint, the elastic energy per unit volume may be reduced by the formation of polydomain structures (shown in the outer circle of Figure 5.8) consisting of a uniform mixture of two of the three variants. The domains in the polydomain structures are related to each other as twins, i.e., their relative strains:

$$\Delta\varepsilon_{ij} = \varepsilon_i^0 - \varepsilon_j^0 \qquad (5.17)$$

are twinning shear. The twinning shear plane is the common plane for both domains and therefore, the twin domains are compatible and thus stress-free. It means that the strain difference $\Delta\varepsilon_{ij}$ satisfies the equation of compatibility:

$$\mathbf{n} \times \Delta\varepsilon_{ij} \times \mathbf{n} = 0, \qquad (5.18)$$

where $\mathbf{n}$ is a vector normal to the interdomain interface which is a twinning plane. There are two solutions of Equation 5.18 and thus each pair of domains can form two polydomains. For example, the white and gray tetragonal domains with:

$$\Delta \varepsilon_{12} = \varepsilon_1^0 - \varepsilon_2^0 = \begin{pmatrix} e - e' & 0 & 0 \\ 0 & -(e - e') & 0 \\ 0 & 0 & 0 \end{pmatrix} \quad (5.19)$$

may form a polytwin with interfaces along (110) as shown in Figure 5.9 or a polytwin with interfaces along (-110).

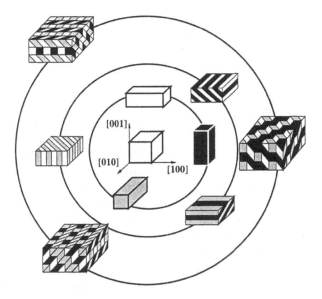

Figure 5.8. Structural domains of the tetragonal lattice (inner circle), two-domain polydomain structures (middle circle), and second-order polydomain structure consisting of all the structural variants of the tetragonal ferroelastic phase (outer circle). The cubic paraelastic phase is shown at the center as a reference (Alpay 2002). Reprinted from "Twinning in Ferroelectric Thin Films: Theory and Structural Analysis". *Handbook of Thin Film Materials, Vol.:3, Ferroelectric and Dielectric Thin Films*, Alpay, S. P., 517-543, Copyright 2000, with permission from Elsevier.

It is clear that simple polydomain structures such as those shown in the middle circle of Figure 5.8 can relax internal stresses only to a certain extent. For complete relaxation, it is necessary to have a polydomain structure that contains all three variants of the tetragonal phase (Alpay and Roytburd 1998; Roytburd, Alpay et al. 2001; Alpay 2002). More hierarchical structures can be realized as twins of twins as shown in the outer circle of Figure 5.8.

Depending on the stress state and the constraint, cellular structures may certainly be possible as well (Slutsker, Armetev et al. 2002).

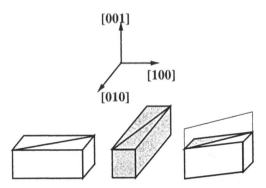

Figure 5.9. The twin relation between the two variants of the tetragonal ferroelastic phase.

The mechanical response of ferroelastics can then be understood in terms of the response of the polydomain structure to an external stress. Since the interdomain interfaces are stress-free, they are extremely mobile in defect-free crystals. Therefore, the structure will adapt itself to external stresses by varying domain fractions. Such a deformation is shown in Figure 5.10 where a uniaxial stress along the [001] direction is applied to the three-domain polydomain crystal. The polydomain structure adapts to the stress via reversible domain wall motion. This motion is accompanied by a self-strain, which may give rise to significant elastic deformation of the crystal depending on the magnitude of the self-strain. The stress-strain response displays a typical hysteretic behavior, Figure 5.11. The origin of the hysteresis is due to the nucleation of white and grey domains at point A. When the stress is reversed one gets nucleation of black domains at point B.

Many technologically important materials and minerals undergo a ferroelastic transformation. The classical example is lead phosphate, $Pb_3(PO_4)_2$. The high-T_C superconductor $YBa_2Cu_3O_{7-x}$ undergoes a tetragonal-to-orthorhombic transformation at 970 K. Associated with this latter transformation is a 2% self-strain. $SrTiO_3$ has a cubic perovskite structure which transforms to a tetragonal lattice via the rotation of the oxygen tetrahedral below 100 K. We refer the reader to the appendices of *Phase Transitions in Ferroelastic and Co-elastic Crystals* (Salje 1990) for an exhaustive list of ferroelastic and co-elastic materials together with the transition temperatures, magnitude of the self-strain and, most importantly, the appropriate forms of the Landau potentials.

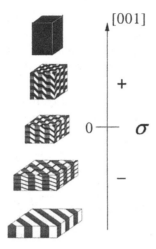

Figure 5.10. The deformation of the three-domain polydomain structure due to an applied uniaxial stress along the [001] direction (Alpay 2002). Reprinted from "Twinning in Ferroelectric Thin Films: Theory and Structural Analysis", *Handbook of Thin Film Materials, Vol.:3, Ferroelectric and Dielectric Thin Films*, Alpay, S. P., 517-543, Copyright 2000, with permission from Elsevier.

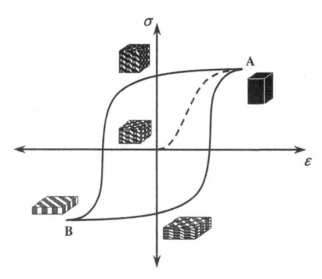

Figure 5.11. A schematic stress-strain curve of a three-domain polydomain structure. The applied stress is along the [001] direction.

Martensitic phase transformations in many metals and alloys can be classified as ferroelastic. The martensitic phase transformation is a diffusion-less, first-order, structural transformation (Nishiyama 1978; Bhattacharya

2003). In fact, Roitburd characterizes the martensitic phase transformation as a typical polymorphic transformation, often accompanied with the formation of a polydomain structure quite similar to ferroelastic oxides (Roitburd 1978; Roitburd and Kurdjumov 1979). The polydomain state may consist of the initial austenitic phase and many orientational variants of the martensitic phase, all related thorough the compatibility relation 5.15. In many alloys, including plain-carbon steels, unlike ferroelastic oxides, the mechanical deformation is not controlled by reversible domain wall motions but rather by the motion of dislocations (of which there are plenty in metallic systems compared to ceramics). Furthermore, the transformational strain may be relaxed by the formation of equilibrium dislocations at the interphase boundaries. However, in certain Ni-Ti, Cu, and Fe-based alloys, the predominant mode of deformation is by the movement of the domain walls as to adapt to the applied stress. These alloys possess the so-called shape memory effect where the austenitic initial phase is remembered after the alloy is deformed in the martensitic state and re-heated (Otsuka 1998). Depending on the vicinity to the phase transformation temperature, the martensitic phase can be induced in the austenitic phase. This stress-induced martensite very often appears as a superelastic deformation due to the high mobility of the interphase interfaces. The material returns to the initial austenitic state during unloading through the same intermediate microstructure states through which it passed under loading (Roytburd and Slutsker 1999). The shape memory effect and superelastic deformation in martensitic alloys is illustrated schematically in Figure 5.12.

The ferroelastic phase transformation can be described via Landau theory (Barsch and Krumhansl 1984; Gooding and Krumhansl 1988; Krumhansl and Gooding 1989; Bales and Gooding 1991; Fradkin 1994; Tang, Zhang et al. 2002). In its simplest form where only a single self-strain is the primary order parameter (a cubic-tetragonal transformation as described by Equations 5.16), the Landau potential takes the form (Krumhansl and Gooding 1989; Bales and Gooding 1991):

$$G(e,T) = G_0 + \frac{1}{2}A(T - T_C)e^2 + \frac{1}{4}Be^4 + \frac{1}{6}Ce^6 \qquad (5.20)$$

where G_0 is the free energy of the paraelastic phase, A, B, and C are the Landau coefficients, and T_C is the phase transformation temperature. The Landau potentials for more complicated ferroelastic phase transformations are summarized by Salje (Salje 1990).

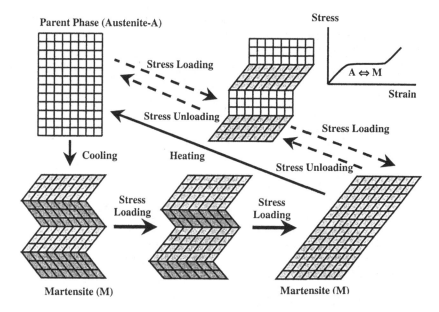

Figure 5.12. The shape memory effect and superelastic deformation in a martensitic alloy. The initial austenite phase (a) transforms to a lower symmetry martensite (b). In the martensitic phase, there are multiple orientational variants. Upon the application of a stress, the domains oriented favorably with respect to the stress grows at the expense of other domains, resulting in a shape change (c). The material "remembers" its shape when heated back to the austenitic phase. A reversible shape change from austenite (a) to martensite (b) can also be induced via an applied stress, the so-called "superelastic" response. A schematic stress-strain curve of superelastic deformation is shown in the inset.

The free energy for a compositionally graded, unconstrained ferroelastic material with systematic composition variations along the direction of the primary order parameter e, has almost an identical form as Equation 2.61 with the notable exception of the electrostatic energy due to depolarization:

$$G = \int_0^L \left[F_0 + \frac{1}{2} A(z)(T - T_C) e^2 + \frac{1}{4} B(z) e^4 + \frac{1}{6} C(z) e^6 \right.$$

$$\left. + \frac{1}{2} D(z) \left(\frac{de}{dz} \right)^2 + F_{el}(z) \right] dz \tag{5.21}$$

where D is the Ginzburg coefficient. In fact, the ferroelastic phase transformation can be realized to be a partial case of the ferroelectric phase transformation where the polarization is the primary order parameter. The ionic displacement also results in a self-strain (Equation 2.4), which can be considered as a secondary order parameter.

The compositional gradient *de/dz* results in a "built-in" stress field and bending. The elastic energy can again be determined form the conditions that the average internal stress and the average momentum of the internal stress should be zero. The internal elastic energy is given by Equation 2.59:

$$F_{el}(z) = \overline{C}\{e(z) - <e> + [z - (L/2)]\kappa\}^2 \qquad (5.22)$$

where *<e>* is the average self-strain.

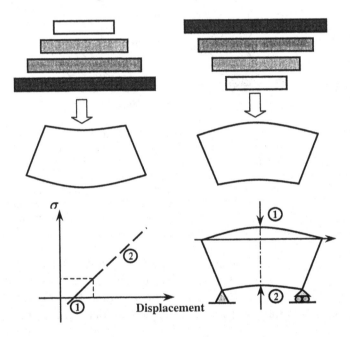

Figure 5.13. Compositionally graded ferroeleastic plate is bent depending on the direction of the composition gradient. The bending angle is highly exaggerated.

The internal elastic field and the resultant bending may be quite significant depending on the magnitude of the self-strain. The displacement of a compositionally graded ferroelastic plate depends on the direction of the composition gradient as shown schematically in Figure 5.13. The mechanical response of such a plate in a four-point bending experiment is clearly not symmetrical and depends on which surface the load is applied. Assuming a linear elastic graded ferroelastics with no plasticity, the fracture stress is achieved at much lower displacements if the load is applied opposite to the direction of grading (see Figure 5.13). A compositionally uniform ferroelastic plate behaves similarly if a temperature gradient is set up across its thickness as shown in Figure 5.14. This interesting behavior suggests that a bi-directional thermal actuator/sensor configuration can be obtained in temperature-graded ferroelectrics. Analogous to graded ferroelectric devices the actuation and sensing capabilities of such a trans-elastic or "translastic" device is expected to be significantly amplified compared to standard thermal actuators and sensors, due to the built-in stress. Considering that ferroelectrics are improper ferroelastic materials, the same mechanical effects are expected in graded ferroelectrics as discussed in Chapter 2. Obviously, the additional benefit would be the built-in potential that results from the polarization gradient, which may eliminate the use of transducers in actuation/sensor applications.

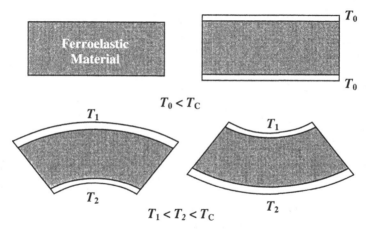

Figure 5.14. A "translastic" device activated by a temperature gradient across the plate thickness. The bending angle is highly exaggerated.

5. SUPERCONDUCTORS

As it has often been noted, Kamerlingh Onnes first observed superconductivity in mercury in 1911, shortly after he successfully liquefied helium (Gorter 1964). Since that time, a host of other materials have displayed similar characteristics, with a great resurgence in the such research coming in 1986 with the discovery of superconductivity in lanthanum based metal oxides by Bednorz and Muller (Bednorz and Muller 1986).

While the most obvious feature of superconductors is their ability to pass current along their surface with zero resistance, other aspects such as the Meissner effect (the exclusion of magnetic flux from the interior of the material) are not a straightforward application of the former property. To address this latter physical property, the phenomenological London equations were developed. This construct related the length over which the superconducting gap parameter can change, referred to as the coherence length ξ, to a fundamental length characterizing the penetration of magnetic flux into the material, referred to as the penetration length λ. When $\xi > \lambda$ the material transitions abruptly from its superconducitive state as the temperature is raised above a critical value, or a sufficiently high magnetic field is applied to the material, Type I superconductors. When $\xi < \lambda$, the transition from the superconducting state is more gradual, with fluxons (line-like regions of non-superconductor) first penetrating the material rather than catastrophic collapse of the superconducting state, Type II superconductors.

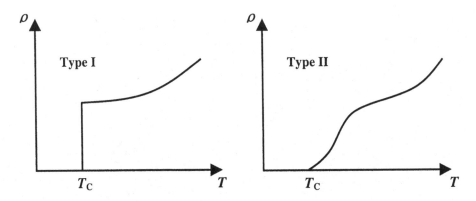

Figure 5.15. Electrical resistivity as a function of temperature for Type I and Type II superconductors.

Ginzburg and Landau (GL) adapted Landau's thermodynamic theory for second-order phase transformations to superconductors (Ginzburg and Landau 1950; Ginzburg 1957). Ginzburg's contribution towards the description of superconductors was recognized in 2003 with a Nobel Prize in physics. While the microscopic Bardeen-Cooper-Schrieffer (BCS) theory (Bardeen, Cooper et al. 1957), Nobel Prize winners in 1972 in physics, provides a more thorough understanding of the physics of type-I superconductors, the GL theory based on the expansion of a quantum (condensate) wave function ψ that serves as the order parameter is obviously simpler and may provide a quite accurate description under certain conditions. It was later shown that the GL model is the continuum limit of the BCS theory and the GL limit can be accurately derived from the BCS model (Gor'kov 1959).

Near the critical temperature T_C, the thermodynamic potential of a superconductor can be expressed in terms of the wave function as:

$$G_s(\psi, T) = G_n + a|\psi|^2 + \frac{b}{2}|\psi|^4 + ... + \frac{h^2}{4m}|\nabla\psi|^2 \qquad (5.23)$$

where G_n is the energy in the normal state, $a=a^*(T-T_C)$ and b are the expansion coefficients, h is the reduced Planck constant, and m is the mass of an electron. The gradient term is the additional energy due to non-uniform order parameter. The equilibrium is given by the equation of the state

$$\frac{\partial G_s}{\partial \psi} = 0 \qquad (5.24)$$

In the presence of an applied magnetic field, the energy of the magnetic field $H^2/2\mu_0$, where μ_0 is the permeability of free space, has to be included in the potential. Furthermore, recalling that superconductivity is due to the flow of Cooper electron pairs, their kinetic energy has to be incorporated as well. Defining the magnetic flux density (or induction) as:

$$\mathbf{B} = \mu\mu_0\mathbf{H}, \qquad (5.25)$$

a vector potential $\mathbf{A}$:

$$\mathbf{B} = \nabla \times \mathbf{A}, \qquad (5.26)$$

and employing the gauge invariance principle (Aitchison and Hey 1996), the effect of the interaction of the Cooper pairs with the external magnetic field can be incorporated into the GL potential, yielding:

$$G_s = G_n + a|\psi|^2 + \frac{b}{2}|\psi|^4 + \frac{h^2}{4m}\left|\left(\nabla - \frac{2ie}{hc}\mathbf{A}\right)\psi\right|^2 + \frac{H^2}{2\mu_0} \qquad (5.27)$$

where e is the electronic charge and c is the speed of light. The free energy is gauge invariant, i.e., the transformation $\mathbf{A}\rightarrow\mathbf{A}+\nabla\Gamma$, where Γ is a scalar field, yields the same free energy if $\psi\rightarrow\psi\,exp(-ie\Gamma/c)$ (Aitchison and Hey 1996). The minimization of the free energy with respect to the order parameter and the electromagnetic vector potential results the GL equations:

$$a\psi + b|\psi|^2\,\psi - \frac{h^2}{4m}\left(\nabla\psi - \frac{2ie}{hc}\mathbf{A}\right)^2\psi = 0 \qquad (5.28)$$

$$\mathbf{J} = -\frac{ieh}{2m}(\psi*\nabla\psi - \psi\nabla\psi*) - \frac{2e^2}{mc}|\psi|\mathbf{A} \qquad (5.29)$$

where $\mathbf{J}$ is the electrical current and ψ^* is the complex conjugate of ψ. We refer the reader to Dmitriev and Nolting's excellent review of the nuances of the GL approach (Dmitriev and Nolting 2004).

Non-uniformities in the superconducting order parameter, ψ- the pairing function, can be created by a number of methods: the introduction of normal materials or even insulators between nearly identical superconductors, subjecting segments of the superconducting elements to a magnetic field, or the imposition of an externally applied voltage across a normal junction between the superconducting elements. In all cases, the effective order parameter variation becomes a phase difference along the length of the non-uniformity.

When the phase difference is a constant, the DC Josephson effect results, that is a persistent non-zero current across a normal junction. When the phase difference increases linearly with time, such as with the imposition of a DC voltage across a normal junction, the AC Josephson effect results, i.e., an oscillatory current across the junction. In essence, these effects are a direct manifestation of a non-constant order parameter and have yielded a host of superconducting quantum interference devices (SQUIDs), including: magnetometers, oscillators, mixers, filters, and amplifiers (Lounasmaa 1974). Clearly, the superconducting transponent devices have developed quite independently of transpacitors and therefore have its own nomenclature and methodologies that are germane to such structures.

6.   MULTI-FERROICS

The term multi-ferroic is almost exclusively employed to describe materials systems that display both ferroelectric and ferromagnetic properties. However, we note that a more suitable term for such materials would be ferro-magnetoelectrics or simply magnetoelectrics since considering that ferroelectrics and ferromagnetic materials are already multi-ferroic systems due to the structural variations during a ferroelectric and ferromagnetic phase transformation. The structural and ferroelectric/ferromagnetic transformation occur at the same temperature and both transformations are, therefore, improper ferroelastic phase transformations. However, there are several other materials systems where the material undergoes successive structural and a ferromagnetic or ferroelectric transformation. For example, $SrTiO_3$ undergoes a structural transformation due to the tilting of oxygen octahedral around 105K and a ferroelectric transformation can be induced via epitaxial stresses at much lower temperatures. In certain ferromagnetic alloys that also display a reversible martensitic transformation, a shape-memory effect can be realized (James and Wuttig 1998; Buchelnikov, Romanov et al. 1999; Murray, Marioni et al. 2000; Wuttig, Liu et al. 2000; Craciunescu and Wuttig 2003; Kreissl, Neumann et al. 2003). Usually the martensitic phase transformation is lower than the Curie temperature. In these alloys, interrelated magnetic ordering and a structural phase transformation resulting in a polydomain structure lead to multi-ferroic properties. A ferromagnetic shape memory effect can be found in several Fe-based alloys in the Fe-Pt, Fe-Pd, Fe-Ni-Co-Ti systems, and in Heusler alloys Ni_2MnGa and Cu_2MnSb (Wuttig, Liu et al. 2000).

Magnetoelectrics have gained considerable importance over the past years due to the co-existence of a spontaneous polarization and magnetization. Materials that display this effect are rare and the spontaneous polarization and/or the magnetization are usually small. Single-phase oxide materials exhibiting coexistence of ferromagnetism and ferroelectricity are limited to a few perovskite-type oxides $BiFeO_3$ or $BiMnO_3$, some rare-earth manganites and molybdates (Goshen, Mukamel et al. 1969; Popov, Kadomtseva et al. 2000; Katsufuji, Mori et al. 2001; Fiebig, Lottermoser et al. 2002; Kimura, Goto et al. 2003; Kimura, Kawamoto et al. 2003; Wang, Neaton et al. 2003). Magnetoelectricity has also been reported in $TbMn_2O_5$ (Saito and Kohn 1995; Hur, Park et al. 2004) and very recently in ultra-thin $SrBi_2Ta_2O_{11}$ films (Tsai 2003).

The Landau potential of two interacting one-dimensional order parameters polarization and magnetization of a magnetoelectric material can be expressed as (Salje 1990):

$$G(P,M,T) = \frac{1}{2}A_1(T - T_{C,1})P^2 + \frac{1}{4}B_1P^4 + \dots$$
$$+ \frac{1}{2}A_2(T - T_{C,2})M^2 + \frac{1}{4}B_2M^4 + \dots \quad (5.30)$$
$$+ \omega_1 P \cdot M + \omega_2 P^2 \cdot M^2$$

where $T_{C,1}$ and $T_{C,2}$ are the temperature of the ferroelectric and ferromagnetic phase transformation, respectively, and ω_1 and ω_2 are linear and bilinear magnetoelectric coupling coefficients. For small (and positive) ω_1 and ω_2, the polarization and magnetization in the "sublattices" hardly interact. For materials that display magnetoelectricity, ω_1 and ω_2 should be large and positive. The spontaneous polarization and magnetization is given through the (coupled) equations of state:

$$\frac{\partial G}{\partial M} = 0, \quad \frac{\partial G}{\partial P} = 0 \qquad (5.31)$$

An interaction between a ferroelectric and a ferromagnetic material can also be established through strain coupling of the polarization and magnetization. Remarkable magnetoelectric properties have been measured for (Pb,Zr)TiO$_3$–Terfenol-D (Tb-Dy-Fe alloy) laminated composites (Cai, Zhai et al. 2003). Recently, it was shown that nanopillars of spinel CoFe$_2$O$_4$ embedded in epitaxial BaTiO$_3$ matrix also display magnetoelectric properties as shown in Figure 5.15 (Zheng, Wang et al. 2004). Magnetoelectric coupling was also achieved artificially in composition spreads ferroelectric BaTiO$_3$, and piezomagnetic CoFe$_2$O$_4$ (Chang, Aronova et al. 2004). The coupling in these composite nanostructures was attributed to strong elastic interactions. The thermodynamic potential for this case can be constructed as:

$$G(P,M,T,f) = (1-f) \cdot [F_P(P,T) - E \cdot P] + f \cdot [F_P(M,T) - H \cdot M] \\ + F_{el}[e(P), e(M)] \qquad (5.32)$$

where

$$F_P(P,T) = F_{0,1} + \frac{1}{2} A_1 (T - T_{C,1}) P^2 + \frac{1}{4} B_1 P^2 + ...$$

$$F_M(M,T) = F_{0,2} + \frac{1}{2} A_2 (T - T_{C,2}) M^2 + \frac{1}{4} B_2 M^2 + ... \qquad (5.33)$$

and f is the volume fraction of the ferromagnetic material, and $e(P)$ and $e(M)$ are the spontaneous ferroelectric and ferromagnetic strain, respectively.

The coupling coefficients of multi-ferroics, ω_1 and ω_2, potentially permit three levels of excitation and active operation. (1) The grading of ferroic phase 1 and its active operation with sensing in ferroic 2. (2) The inverse case, where ferroic 2 is compositionally graded or there exist a natural grading in order parameter, with sensing in ferroic 1. (3) Some combination of (1) and (2). For the case of magneto-electric transponents, one can thus imagine active: electromagnetic wave lens, tuners, circulators, and phase variable or steerable antenna systems. Multi-ferroics, therefore, inherently multiply expand the range of potential transponent systems.

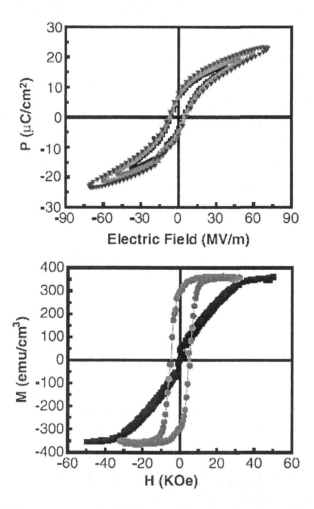

Figure 5.15. (a) Polarization-electric field hysteresis loop showing that the film is ferroelectric with a saturation polarization $P_S \sim 23$ $\mu C/cm^2$. (b) Out-of-plane (gray) and in plane (black) magnetic hysteresis loops depicting the large uniaxial anisotropy (Zheng, Wang et al. 2004). Reprinted with permission from [Zheng, H., Wang, J., Lofland, S. E., Ma, Z., Mohaddes-Ardabili, L., Zhao, T., Salamanca-Riba, L., Shinde, S. R., Ogale, S. B., Bai, F., Viehland, D., Jia, Y., Schlom, D. G., Wuttig, M., Roytburd, A. L. and Ramesh, R., "Multiferroic BaTiO$_3$-CoFe$_2$O$_4$ Nanostructures," *Science* **303**, 661-663, (2004)]. Copyright 2004 AAAS.

We note that in several high-T_C superconducting complex oxides, including many in the LaCuO (Weidinger, Niedermayer et al. 1989; Ostenson,

Bud'ko et al. 1997; Tranquada, Axe et al. 1997; Nachumi, Fudamoto et al. 1998; Niedermayer, Bernhard et al. 1998; Suzuki, Goto et al. 1998; Yamada, Lee et al. 1998; Hunt, Singer et al. 1999; Curro, Hammel et al. 2000; Ichikawa, Uchida et al. 2000) and Ru-based systems (Wu, Chen et al. 1997; Knigavko, Rosenstein et al. 1999; Cao, Xin et al. 2001; Felner 2002; Papageorgiou, Herrmannsdorfer et al. 2002; Kuzmin, Ovchinnikov et al. 2003), the co-existence of magnetism and superconductivity has been reported. The free energy of these systems with coupled magnetic and superconductor ordering can be expressed quite similar to a magnetoelectric system (Equation 5.32) as (Kivelson, Aeppli et al. 2001; Kivelson, Lee et al. 2002):

$$G(\psi, M, T) = G_s(\psi, T) + G_M[M(\mathbf{r}), T] + \omega \cdot |\psi| \cdot |M(\mathbf{r})|^2 \quad (5.34)$$

where

$$G_M[M(\mathbf{r}), T] = G_0 + \frac{1}{2} A(T - T_C) M(\mathbf{r})^2 + \frac{1}{4} BM(\mathbf{r})^4$$
$$+ \frac{1}{2} D[\nabla M(\mathbf{r}) \cdot \nabla M(\mathbf{r})] \quad (5.35)$$

$$G_s(\psi, T) = G_n + a|\psi|^2 + \frac{b}{2}|\psi|^4 + \dots + \frac{h^2}{4m}|\nabla \psi|^2 \quad (5.36)$$

and ω is a bilinear coupling coefficient. For a strong interaction between the order parameters, ω is positive and large. For small ω, the order parameters are decoupled and act independently from each other.

7.   OTHER FUNCTIONAL AND SMART MATERIALS SYSTEMS

Wang and Kang (Wang and Kang 1998) define functionally "smart" materials as being distinctly different from structural materials with "...*physical and chemical properties that are sensitive to a change in the environment such as temperature, pressure, electric field, magnetic field, optical wavelength, adsorbed gas molecules, and pH.*"

Examples cover the whole spectrum of materials systems in solid-state physics including ferroelectrics, ferro- and ferrimagnets, semiconductors, superconductors, photonic crystals, and many transition metal oxides with metal-insulator transitions. As discussed in great detail in the preceding chapters, ferroelectrics have unique properties. To name a few, they can sense changes in temperature, electric field, and the stress state. They can actuate due to their piezoelectric properties. Similarly, ferromagnetic (and ferrimagnetic) materials can be employed for the very same purposes where the sensing and actuating is accomplished by magnetic stimulation/response. Photonic semiconductors are based on excitations of electrons across the band gap. Electrons excited to higher energy levels emit photons with shorted wavelengths than electrons excited to lower levels. Excited electrons can resume their normal level spontaneously or a photon with the proper wavelength can stimulate an excited electron to return to its normal level. Energy entering a semiconductor crystal excites electrons to higher levels, leaving behind "holes". These electrons and "holes" can recombine and emit photons, or they can move away from one another and form a current. This is essentially the basics of semiconductor light detectors. Light Emitting Diodes (LEDs) based on GaAs and SiC convert electrical current directly into light. In many ways, the LED is more efficient than many other light sources. Superconductors are elements (Hg, W), intermetallic alloys (Nb_3Sn), or complex ceramic compounds ($YBa_2Cu_3O_{7-x}$) that conduct electricity without any resistance below a certain critical temperature. Certain transition metal oxides or solid solutions behave in a similar manner as they are cooled down through a critical temperature (Rao 1998) although this effect is not as dramatic as superconductors. For example, certain composition ranges in the solid solutions of $La_{1-x}Ca_xMnO_3$ and $La_{1-x}Sr_xMnO_3$ undergo a paramagnetic insulator to ferromagnetic metal transition below a critical temperature. There is orders of magnitude increase in the electrical conductivity associated with this metal-insulator transition. Another classical example is V_2O_3. It undergoes a first-order transformation (monoclinic-rhombohedral) at $-123°C$ accompanied by a seven orders of magnitude drop in the resistivity. There are a great many organic smart material systems including conducting and shape-memory polymers.

It is therefore logical to also ask, "Is it possible to extend our discussion of graded ferroics yet further, this time taking into account graded smart materials which can be characterized by an order parameter and associated conjugate field?" Such a question is intriguing. A partial survey of the many possible smart material systems that fit such a description is given in Table 5.3 (Mazenko 2002), though such a list is by no means exhaustive.

It is worth recalling at this time, that the study of polarization-graded ferroelectrics (alone) has been ongoing for more than ten years. Yet this latter ferroic system still continues to yield new experimental and theoretical characteristics as the inner complexity of the devices reveal themselves.

Moreover, the graded ferroelectric and transpacitive configurations studied to date can only be described as rudimentary; lacking the sophisticated constructs of their more advanced semiconductor counterparts. Indeed, as of this writing, there does not even exist an adequate theory or even detailed experimental results that can adequately describe the external field dependence of transpacitors. Therefore, we do not (at this time) take up the task posed above, but suggest it as an area of future research that may benefit and be addressed using some of the tools that we have developed for graded ferroics, most particularly graded ferroelectrics.

Table 5.3. Order parameters and conjugate fields for various systems.

System	Order Parameters	Conjugate Field
Superfluid ^{4}He	Creation Operator	Particle Source
Phase separation in Alloys	Density of Components	Chemical Potential
Order disorder in Alloys	Sublattice Density of Components	Chemical Potential
Potts Model	Permutation Density	Break Permutation Symmetry
Nematic Liquid Crystal	Traceless Director Tensor	Electric Field
Smectic *A* Liquid Crystal	Layer Density	Magnetic field
Phase Separation in Complex Fluids	Relative Concentration	Difference in Chemical Potentials

References

Aitchison, I. J. R. and Hey, A. J. G., Gauge Theories in Particle Physics, Institute of Physics Publishing (1996).

Aizu, K., "Possible Species of Ferromagnetic, Ferroelectric, and Ferroelastic Crystals," Physical Review B 2, 754-772, (1970).

Alpay, S. P. "Twinning in Ferroelectric Thin Films: Theory and Structural Analysis". Handbook of Thin Film Materials, Vol.:3, Ferroelectric and Dielectric Thin Films. H. S. Nalwa. San Diego, Academic Press: 517-543. (2002)

Alpay, S. P., Ban, Z.-G. and Mantese, J. V., "Thermodynamic Analysis of Temperature-Graded Ferroelectrics," Applied Physics Letters 82, 1269, (2003).

Alpay, S. P. and Roytburd, A. L., "Thermodynamics of Polydomain Heterostructures. III. Domain Stability Map," Journal of Applied Physics 83, 4714-4723, (1998).

Anderson, J. C., Dielectrics, London, Chapman and Hall Ltd. (1964).

Ashcroft, N. W. and Mermin, N. D., Solid State Physics, New York, Saunders College Publishing (1976).

Bales, G. S. and Gooding, R. J., "Interfacial Dynamics at a First-order Phase Transition Involving Strain: Dynamical Twin Formation," Physical Review Letters 67, 3412-3415, (1991).

Ban, Z.-G., Alpay, S. P. and Mantese, J. V., "Fundamentals of Graded Ferroic Materials and Devices," Physical Review B 67, 184104, (2003).

Ban, Z.-G., Alpay, S. P. and Mantese, J. V., "Hysteresis Offset and Dielectric Response of Compositionally Graded Ferroelectric Materials," Integrated Ferroelectrics 58, 1281-1291, (2003).

Bao, D., Mizutani, N., Yao, X. and Zhang, L., "Dielectric and Ferroelectric Properties of Compositionally Graded (Pb, La)TiO₃ Thin Films on Pt/Ti/SiO₂/Si Substrates," Applied Physics Letters 77, 1203-1205, (2000).

Bao, D., Mizutani, N., Yao, X. and Zhang, L., "Structural, Dielectric, and Ferroelectric Properties of Compositionally Graded (Pb,La)TiO₃ Thin Films with Conductive LaNiO₃ Bottom Electrodes," Applied Physics Letters 77, 1041-1043, (2000).

Bao, D., Mizutani, N., Zhang, L. and Yao, X., "Compositional Gradient Optimization and Electrical Characterization of (Pb,Ca)TiO₃ Thin Films," Journal of Applied Physics 89, 801-803, (2001).

Bao, D., Wakiya, N., Shinozaki, K., Mizutani, N. and Yao, X., "Abnormal Ferroelectric Properties of Compositional Graded Pb(Zr,Ti)O₃ Thin Films with LaNiO₃ Bottom Electrodes," Journal of Applied Physics 90, 506-508, (2001).

Bao, D., Yao, X. and Zhang, L., "Dielectric Enhancement and Ferroelectric Anomaly of Compositionally Graded (Pb,Ca)TiO₃ Thin Films Derived by a Modified Sol-Gel Technique," Applied Physics Letters 76, 2779-2781, (2000).

138

Bao, D., Zhang, L. and Yao, X., "Compositionally Step-Varied (Pb, Ca)TiO$_3$ Thin Films with Enhanced Dielectric and Ferroelectric Properties," Applied Physics Letters 76, 1063-1065, (2000).

Bardeen, J., Cooper, L. N. and Schrieffer, J. R., "Theory of Superconductivity," Physical Review 108, 1175, (1957).

Barsch, G. R. and Krumhansl, J. A., "Twin Boundaries in Ferroelastic Media without Interface Dislocations," Physical Review Letters 53, 1069-1072, (1984).

Bartic, A. T., Wouters, D. J., Maes, H. E., Rickes, J. T. and Waser, R. M., "Preisach Model for the Simulation of Ferroelectric Capacitors," Journal of Applied Physics 89, 3420-3425, (2001).

Bednorz, J. G. and Muller, K. A., "Possible High T$_C$ Superconductivity in the Ba-La-Cu-O System," Zeitschrift Fur Physik B 64, 189-193, (1986).

Bhattacharya, K., Microstructure of Martensite: Why it forms and how it Gives Rise to the Shape-Memory Effect, Oxford, Oxford University Press (2003).

Boerasu, I., Pintilie, L. and Kosec, M., "Ferroelectric Properties of Pb$_{1-3y/2}$La$_y$(Zr$_{0.4}$Ti$_{0.6}$)O$_3$ Structures with La Concentration Gradients," Applied Physics Letters 77, 2231, (2000).

Bozoroth, R. M., Ferromagnetism, Pitcataway, IEEE Press (1978).

Bratkovsky, A. M. and Levanyuk, A. P., "Easy Collective Polarization Switching in Ferroelectrics," Physical Review Letters 85, 4614-4617, (2000).

Bratkovsky, A. M. and Levanyuk, A. P., "Formation and Rapid Evolution of Domain Structure at Phase Transitions in Slightly Inhomogenous Ferroelectrics and Ferroelastics," Physical Review B 66, 184109, (2002).

Brazier, M. and McElfresh, M., "Origin of Anomalous Polarization Offsets in Compositionally Graded Pb(Zr,Ti)O$_3$ Thin Films," Applied Physics Letters 74, 299-301, (1999).

Brazier, M., McElfresh, M. and Mansour, S., "Unconventional Hysteresis Behavior in Compositionally Graded Pb(Zr,Ti)O$_3$ Thin Films," Applied Physics Letters 72, 1121-1123, (1998).

Buchelnikov, V. D., Romanov, V. S. and Zayak, A. T., "Structural Phase Transitions in Cubic Ferromagnets," Journal of Magnetism and Magnetic Materials 191, 203-206, (1999).

Cai, N., Zhai, J., Nan, C. W., Lin, Y. and Shi, Z., "Dielectric, Ferroelectric, Magnetic, and Magnetoelectric Properties of Multiferroic Laminated Composites," Physical Review B 68, 224103, (2003).

Canedy, C. L., Li, H., Alpay, S. P., Salamanca-Riba, L., Roytburd, A. L. and Ramesh, R., "Dielectric Properties in Heteroepitaxial Ba$_{0.6}$Sr$_{0.4}$TiO$_3$ Thin Films: Effect of Internal Stresses and Dislocation-type Defects," Applied Physics Letters 77, 1695-1697, (2000).

Cao, G., Xin, Y., Alexander, C. S. and Crow, J. E., "Weak Ferromagnetism and Spin-charge Coupling in Single Crystal Sr$_2$YRuO$_6$," Physical Review B 63, 184432, (2001).

Cao, H. X. and Li, Z. Y., "Thermodynamic Properties of Temperature Graded Ferroelectric Film," Journal of Physics: Condensed Matter 15, 6301-6310, (2003).

Chai, F. K., Brews, J. R., Schrimpf, R. D. and Birnie III, D. P., "Domain Switching and Spatial Dependence of Permittivity in Ferroelectric Thin Films," Journal of Applied Physics 82, 2505-2516, (1997).

Chan, H. K., Lam, C. H. and Shin, F. G., "Time-dependent Space-charge-limited Conduction as a Possible Origin of the Polarization Offsets Observed in Compositionally Graded Ferroelectric Films," Journal of Applied Physics 95, 2665, (2004).

Chang, K. S., Aronova, M. A., Lin, C. L., Murakami, M., Yu, M. H., Hattrick-Simpers, J., Famodu, O. O., Lee, S. Y., Ramesh, R., Wuttig, M., Takeuchi, I., Gao, C. and Bendersky, L. A., "Exploration of Artificial Multiferroic Thin-film Heterostructures Using Composition Spreads," Applied Physics Letters 84, 3091-3093, (2004).

Chen, T., Malmhall, R. and Charlan, G. B., "Anomalous Hysteresis Loops in Single and Double Layer Sputtered TbFe Films," Journal of Magnetism and Magnetic Materials 35, 269-271, (1983).

Chen, W. and Lynch, C. S., "A Micro-electro-mechanical Model for Polarization Switching of Ferroelectric Materials," Acta Materialia 46, 5303-5311, (1998).

Chen, X., Fang, D. N. and Hwang, K. C., "Micromechanics Simulation of Ferroelectric Polarization Switching," Acta Materialia 45, 3181-3189, (1997).

Chen, Z., Arita, K., Lim, M. and Araujo, C. A. P., "Graded PZT Thin Film Capacitors with Stoichimetric Variation by MOD Technique," Integrated Ferroelectrics 24, 181-188, (1999).

Chew, K., Ishibashi, Y., Shin, F. G. and Chan, H. L.-W., "Theory of Interface Structures in Double-Layer Ferroelectrics," Journal of the Physical Society of Japan 72, 2364-2368, (2003).

Christian, J. W., The Theory of Transformations in Metals and Alloys, Part 1, Equilibrium and General Kinetic Theory, New York, Pergamon Press (1975).

Cottam, M. G., Tilley, D. R. and Žeks, B., "Theory of Surface modes in Ferroelectrics," Journal of Physics C: Solid State Physics, (1984).

Craciunescu, C. M. and Wuttig, M., "New Ferromagnetic and Functionally Graded Shape Memory Alloys," Journal of Optoelectronics and Advanced Materials 5, 139-146, (2003).

Cross, L. E., "Relaxor Ferroelectrics," Ferroelectrics 76, 241-267, (1987).

Cross, L. E., "Relaxor Ferroelectrics: An Overview," Ferroelectrics 151, 305-320, (1994).

Curro, N. J., Hammel, P. C., Suh, B. J., Hucker, M., Buchner, B., Ammerahl, U. and Revcolevschi, A., "Inhomogeneous Low Frequency Spin Dynamics in $La_{1.65}Eu_{0.2}Sr_{0.15}CuO_4$," Physical Review Letters 85, 642-645, (2000).

Damjanovic, D., "Ferroelectric, Dielectric and Piezoelectric Properties of Ferroelectric Thin Films and Ceramics," Reports on Progress in Physics 61, 1267-1324, (1998).

de Fontaine, D. "Configurational Thermodynamics of Solid Solutions". Solid State Physics. H. Ehrenreich, F. Seitz and D. Turnbull. New York, Academic Press Inc. 34: 73. (1979)

Devonshire, A. F., "Theory of Barium Titanate - Part I.," Philosophical Magazine 40, 1040-1063, (1949).

Devonshire, A. F., "Theory of Barium Titanate - Part II," Philosophical Magazine 42, 1065-1079, (1951).

Devonshire, A. F., "Theory of Ferroelectrics," Advances in Physics 3, 85-130, (1954).

Dmitriev, A. V. and Nolting, W., "Details of the Thermodynamical Derivation of the Ginzburg-Landau Equations," Superconductor Science and Technology 17, 443-447, (2004).

Drougard, M. E. and Landauer, R., "On the Dependance of the Switching Time of Barium Titanate Crystals on their Thickness," Journal of Applied Physics 30, 1663-1668, (1959).

Ebers, J. J. and Moll, J. L., "Large-signal behavior of junction transistors," Proceedings of the IRE 42, 1761-1772, (1954).

Erbil, A., Kim, Y. and Gerhardt, R. A., "Giant Permittivity in Epitaxial Ferroelectric Heterostructures," Physical Review Letters 77, 1628-1631, (1996).

Fatuzzo, E. and Merz, W. J., Ferroelectricity, New York, John Wiley and Sons, Inc. (1967).

Fellberg, W., Mantese, J. V., Schubring, N. W. and Micheli, A. L., "Origin of the "Up", "Down" Hysteresis Offsets Observed from Polarization-Graded Ferroelectric Materials," Applied Physics Letters 78, 524-526, (2001).

Felner, I., "Anomalous Magnetic Behavior in $YSr_2(Ru_{0.8}Cu_{0.2})O_6$," Journal of Magnetism and Magnetic Materials 246, 191-198, (2002).

Fiebig, M., Lottermoser, T., Frohlich, D., Goltsev, A. V. and Pisarev, R. V., "Observation of Coupled Magnetic and Electric Domains," Nature 419, 818-820, (2002).

Forsbergh, P. W., Jr., "Domain Structures and Phase Transition in Barium Titanate," Physical Review 76, 1187-1201, (1949).

140

Fradkin, M. A., "External Field in the Landau Theory of a Weakly Discontinuous Phase Transition: Pressure Effect in the Martensitic Transitions," Physical Review B 50, 16326-16338, (1994).

Freund, L. B., "Dislocation Mechanisms of Relaxation in Strained Epitaxial Films," MRS Bulletin 17, 52, (1992).

Freund, L. B., "Some Elementary Connections Between Curvature and Mismatch Strain in Compositionally Graded Thin Films," Journal of the Mechanics and Physics of Solids 44, 723-736, (1996).

Fridkin, V. M., Ferroelectric Semiconductors, New York, Consultants Bureau (1980).

Ginzburg, V. L., "On the Dielectric Properties of Ferroelectric (Segnetoelectric Crystals and Barium Titanate)," Journal of Experimental and Theoretical Physics 15, 1945, (1945).

Ginzburg, V. L., Journal of Experimental and Theoretical Physics 32, 1442, (1957).

Ginzburg, V. L., "Some Remarks on Ferroelectricity, Soft Modes and Related Problems," Ferroelectrics 76, 3-22, (1987).

Ginzburg, V. L. and Landau, L. D., Journal of Experimental and Theoretical Physics 20, 1064, (1950).

Glinchuk, M. D., Eliseev, E. A. and Stephanovich, V. A., "The Depolarization Field Effect on the Thin Ferroelectric Films Properties," Physica B 322, 356-370, (2002).

Glinchuk, M. D., Eliseev, E. A., Stephanovich, V. A. and Farhi, R., "Ferroelectric Thin Film Properties-Depolarization Field and Renormalization of a "Bulk" Free Energy Coefficients," Journal of Applied Physics 93, 1150-1159, (2003).

Gooding, R. J. and Krumhansl, J. A., "Theory of Bcc-to-9R Structural Phase Transformation of Li," Physical Review B 38, 1695-1704, (1988).

Gor'kov, L. P., Journal of Experimental and Theoretical Physics 36, 1364, (1959).

Gorter, C. J., "Superconductivity until 1940 In Leiden and As Seen From There," Reviews of Modern Physics 36, 3-7, (1964).

Goshen, S., Mukamel, D. and H., S., "Magnetic Symmetry and Ferroelectricity Induced by Antiferromagnetic Transitions," Journal of Applied Physics 40, 1590-1592, (1969).

Guttman, L. "Order-Disorder Phenomena in Metals". Solid State Physics: Advances in Research and Applications. New York, Academic Press Inc.: 146. (1956)

Harrison, R. J. "Magnetic Transitions in Minerals". Reviews in Mineralogy and Geochemistry Vol. 39: Transformation Processes in Minerals. Washington, D.C., GMSA. (2000)

Haun, M. J., Furman, E., Jang, S. J. and Cross, L. E., "Thermodynamic Theory of The Lead Zirconate-Titanate Solid Solution System," Ferroelectrics 99, 13-86, (1989).

Haun, M. J., Furman, E., Jang, S. J. and Cross, L. E., "Thermodynamic Theory of The Lead Zirconate-Titanate Solid Solution System, Part II: Tricritical Behavior," Ferroelectrics 99, 27-44, (1989).

Haun, M. J., Furman, E., Jang, S. J. and Cross, L. E., "Thermodynamic Theory of The Lead Zirconate-Titanate Solid Solution System, Part III: Curie Constant and Sixth-Order Polarization Interaction Dielectric Stiffness Coeffcients," Ferroelectrics 99, 45-54, (1989).

Haun, M. J., Furman, E., Jang, S. J. and Cross, L. E., "Thermodynamic Theory of The Lead Zirconate-Titanate Solid Solution System, Part IV: Tilting of The Oxygen Octahedra," Ferroelectrics 99, 55-62, (1989).

Haun, M. J., Furman, E., Jang, S. J. and Cross, L. E., "Thermodynamic Theory of The Lead Zirconate-Titanate Solid Solution System, Part V: Theoritical Calculations," Ferroelectrics 99, 63-86, (1989).

Haun, M. J., Furman, E., Jang, S. J., McKinstry, H. A. and Cross, L. E., "Thermodynamic Theory of $PbTiO_3$," Journal of Applied Physics 62, 3331, (1987).

Hilton, A. D. and Ricketts, B. W., "Dielectric Properties of $Ba_{1-x}Sr_xTiO_3$," Journal of Physics D: Applied Physics 29, 1321-1325, (1996).

Horowitz, P. and Hill, W., The Art of Electronics, New York, Cambridge University Press (1980).

Hubert, A., Theorie der Domänenwände in geordneten Medien, Berlin, Springer-Verlag (1974).

Hubert, A. and Schafer, R., Magnetic Domains: The Analysis of Magnetic Microstructures, Berllin, Springer-Verlag (1998).

Hummel, R. E., Electronic Properties of Materials, New York, Springer-Verlag, Inc. (2001).

Hunt, A. W., Singer, P. M., Thurber, K. R. and Imai, T., "^{63}Cu NQR Measurement of Stripe Order Parameter in La$_{2-x}$Sr$_x$CuO$_4$," Physical Review Letters 82, 4300-4303, (1999).

Hur, N., Park, S., Sharma, P. A., Ahn, J. S., Guha, S. and Cheong, S.-W., "Electric Polarization Reversal and Memory in Multiferroic Material Induced by Magnetic Fields," Nature 429, 392-395, (2004).

Hwang, S. C., Huber, J. E., McMeeking, R. M. and Fleck, N. A., "The Simulation of Switching in Polycrystalline Ferroelectric Ceramics," Journal of Applied Physics 84, 1530-1540, (1998).

Ichikawa, N., Uchida, S., Tranquada, J. M., Niemoller, T., Gehring, P. M., Lee, S. H. and Schneider, J. R., "Local Magnetic Order vs Superconductivity in a Layered Cuprate," Physical Review Letters 85, 1738-1741, (2000).

Ishibashi, Y., "Theory of Polarization Reversals in Ferroelectric Based on Landau-Type Free Energy," Japanese Journal of Applied Physics 31, 2822-2824, (1992).

Ishibashi, Y. and Orihara, H., "The Characteristics of a Poly-Nuclear Growth Model," Journal of the Physical Society of Japan 55, 2315-2319, (1986).

Ishibashi, Y. and Orihara, H., "Size Effect in Ferroelectric Switching," Journal of the Physical Society of Japan 61, 4650-4656, (1992).

Ishibashi, Y. and Takagi, Y., "Note on Ferroelectric Domain Switching," Journal of the Physical Society of Japan 31, 506-510, (1971).

Jackson, J. D., Classical Electrodynamics, John Wiley and Sons, Inc. (1998).

James, R. D. and Wuttig, M., "Magnetostriction of Martensite," Philosophical Magazine A 77, 1273-1299, (1998).

Jayadevan, K. P. and Tseng, T. Y., "Review: Composite and Multilayer Ferroelectric Thin Films: Processing, Properties and Applications," Journal of Materials Science: Materials in Electronics 13, 439-459, (2002).

Jiang, J. C., Pan, X. Q., Tian, W., Theis, C. D. and Schlom, D. G., "Abrupt PbTiO$_3$/SrTiO$_3$ Superlattices Grown by Reactive Molecular Beam Epitaxy," Applied Physics Letters 74, 2851-2853, (1999).

Jiles, D., Introduction to Magnetism and Magnetic Materials, London, CRC Press (1998).

Jin, F., Auner, G. W., Naik, R., Schubring, N. W., Mantese, J. V. and Micheli, A. L., "Giant Effective Pyroelectric Coefficients from Graded Ferroelectric Devices," Applied Physics Letters 73, 2838-2840, (1998).

Jona, F. and Shirane, G., Ferroelectric Crystals, New York, Dover Publications, Inc. (1962).

Kadanoff, L. P., Gotze, W., Hamblen, D., Hecht, R., Lewis, E. A. S., Paiciauskas, V. V., Rayl, M., Swift, J., Aspnes, D. and Kane, J., "Static Phenomena Near Critical Points: Theory and Experiment," Review of Modern Physics 39, 395-431, (1967).

Kaiser, C. J., Capacitor Handbook, Wiley, John & Sons, Incorporated (March 1993).

Kanno, I., Hayashi, S., Takayama, R. and Hirao, T., "Superlattices of PbZrO$_3$ and PbTiO$_3$ Prepared by Multi-ion-beam Sputtering," Applied Physics Letters 68, 328, (1996).

Katsufuji, T., Mori, S., Masaki, M., Moritomo, Y., Yamamoto, N. and Takagi, H., "Dielectric and Magnetic Anomalies and Spin Frustration in Hexagonal RMnO$_3$ (R=Y, Yb, and Lu)," Physical Review B 64, 104419-1, (2001).

Kim, H. J., Oh, S. H. and Jang, H. M., "Thermodynamic Theory of Stress Distribution in Epitaxial Pb(Zr, Ti)O$_3$ Thin Films," Applied Physics Letters 75, 3195, (1999).

Kim, Y., Gerhardt, R. A. and Erbil, A., "Dynamical Properties of Epitaxial Ferroelectric Superlattices," Physical Review B 55, 8766-8775, (1997).

Kimura, T., Goto, T., Shintani, H., Ishizaka, K., Arima, T. and Yokura, Y., "Magnetic Control of Ferroelectric Polarization," Nature 426, 55-58, (2003).

Kimura, T., Kawamoto, S., Yamada, I., Azuma, M., Takano, M. and Tokura, Y., "Magnetocapacitance Effect in Multiferroic $BiMnO_3$," Physical Review B 67, 180401-1, (2003).

Kittel, C., "Theory of Structure of Ferromagnetic Domains in Films and Small Particles," Physical Review 70, 965-971, (1946).

Kittel, C., Introduction to Solid State Physics, New York, John Wiley and Sons (1996).

Kivelson, S. A., Aeppli, G. and Emery, V. J., "Thermodynamics of the Interplay Between Magnetism and High-temperature Superconductivity," Proceedings of the National Academy of Sciences of the United States of America 98, 11903-11907, (2001).

Kivelson, S. A., Lee, D. H., Fradkin, E. and Oganesyan, V., "Competing Order in the Mixed State of High Temperature Superconductors," Physical Review B 66, 144516, (2002).

Knigavko, A., Rosenstein, B. and Chen, Y. F., "Magnetic Skyrmions and Their Lattices in Triplet Superconductors," Physical Review B 60, 550-558, (1999).

Koebernik, G., Haessler, W., Pantou, R. and Weiss, F., "Thickness Dependence on the Dielectric Properties of $BaTiO_3$/$SrTiO_3$-multilayers," Thin Solid Films 449, 80-85, (2004).

Kooy, C. and Enz, U., "Experimental and Theoritical Study of the Domain Configuration in Thin Layers of $BaFe_{12}O_{19}$," Philips Research Reports 15, 7-29, (1960).

Kostorz (ed.), G., Phase Transformations in Materials, Weinheim, Wiley-VCH (2001).

Kreissl, M., Neumann, K.-U., Stephens, T. and Ziebeck, K. R. A., "The Influence of Atomic Order on the Magnetic and Structural Properties of Ferromagnetic Shape Memory Compound Ni_2MnGa," Journal of Physics: Condensed Matter 15, 3831-3839, (2003).

Kretschmer, R. and Binder, K., "Surface Effects on Phase Transitions in Ferroelectrics and Dipole Magnets," Physical Review B 20, 1065-1075, (1979).

Krumhansl, J. A. and Gooding, R. J., "Structural Phase Transitions with Little Phonon Softening and First-order Character," Physical Review B 39, 3047-3053, (1989).

Kulwicki, B. M., Amin, A., Beratan, H. R. and Hanson, C. M. Ferroelectric Imaging. Proceedings of the 8th International Symposium on Applied Ferroelectrics, New York, IEEE. (1992)

Kuzmin, E. V., Ovchinnikov, S. G. and Singh, D. J., "Effect of Frustrations on Magnetism in the Ru Double Perovskite Sr_2YRuO_6," Physical Review B 68, 024409, (2003).

Landau, L. D. and Lifshitz, E. M., Statistical Physics, Oxford, Pergamon Press (1980).

Levanyuk, A. P., Journal of Experimental and Theoretical Physics 9, 571, (1959).

Li, S., Eastman, J. A., Vetrone, J. M., Newnhan, R. E. and Cross, L. E., "Dielectric Response in Ferroelectric Supperlattices," Philosophical Magazine 76, 47-57, (1997).

Lines, M. E. and Glass, A. M., Principles and Application of Ferroelectrics and Related Materials, Oxford, Clarendon Press (1977).

Lounasmaa, O. V., Experimental Principles and Methods Below 1K, Academic Press (1974).

Ma, Y.-Q., Shen, J. and Xu, X.-H., "Coupling Effects in Ferroelectric Superlattice," Solid State Communications 114, 461-464, (2000).

Mantese, J. V., Micheli, A. L. and Hamdi, A. H., "Metalorganic Deposition (MOD): A Non-vacuum, Spin-on-liquid-based, Thin Film Method," MRS Bulletin 10, (1989).

Mantese, J. V., Schubring, N. W. and Micheli, A. L., "Polarization-Graded Ferroelectrics: Transpacitor Energy Gain," Applied Physics Letters 79, 4007-4009, (2001).

Mantese, J. V., Schubring, N. W. and Micheli, A. L., "Polarization-Graded Ferroelectrics: Transpacitor Push-Pull Amplifer," Applied Physics Letters 80, 1430-1431, (2002).

Mantese, J. V., Schubring, N. W., Micheli, A. L. and Catalan, A. B., "Ferroelectric Thin Films with Polarization Gradients Normal to the Growth Surface," Applied Physics Letters 67, 721-723, (1995).

Mantese, J. V., Schubring, N. W., Micheli, A. L., Mohammed, M. S., Naik, R. and Auner, G. W., "Slater Model Applied to Polarization Graded Ferroelectrics," Applied Physics Letters 71, 2047-2049, (1997).

Mantese, J. V., Schubring, N. W., Micheli, A. L., Thompson, M. P., Naik, R., Auner, G. W., Misirlioglu, I. B. and Alpay, S. P., "Stress Induced Polarization-Graded Ferroelectrics," Applied Physics Letters 81, 1068, (2002).

Marrec, F. L., Farhi, R., Marssi, M. E., Dellis, J. L. and Karjut, M. G., "Ferroelectric $PbTiO_3/BaTiO_3$ Superlattices: Growth Anomalies and Confined Modes," Physical Review B 61, R6447-R6450, (2000).

Marvan, M. and Fousek, J., "Electrostatic Energy of Ferroelectrics with Nonhomogenous Distributions of Polarization and Free Charges," Electrostatic Energy of Ferroelectrics 208, 523-531, (1998).

Matthews, J. W. and Blakeslee, A. E., "Defects in Epitaxial Multilayers: 1. Misfit Dislocation," Journal of Crystal Growth 27, 118, (1974).

Mazenko, G. F., Fluctuations Order & Defects, New Jersey, John Wiley & Sons (2002).

Mendiola, J., Calzada, M. L., Ramos, P., Martin, M. J. and Agullo-Rueda, F., Thin Solid Films 195, 315, (1998).

Merz, W. J., "Domain Formation and Domain Wall Motions in Ferroelectric $BaTiO_3$ Single Crystals," Physical Review 95, 690-698, (1954).

Miller, R. C. and Weinreich, G., "Mechanism for the Sidewise Motion of $180°$ Domain Walls in Barium Titanate," Physical Review 117, 1460-1466, (1960).

Mohammed, M. S., Auner, G. W., Naik, R., Mantese, J. V., Schubring, N. W., Micheli, A. L. and Catalan, A. B., "Temperature Dependence of Conventional and Effective Pyroelectric Coefficients for Compositionally Graded $Ba_xSr_{1-x}TiO_3$ Films," Journal of Applied Physics 84, 3322-3325, (1998).

Murray, S. J., Marioni, M. A., Kukla, A. M., Robinson, J., O'Handley, R. C. and Allen, S. M., "Large Field Induced Strain in Single Crystalline Ni-Mn-Ga Ferromagnetic Shape Memory Alloy," Journal of Applied Physics 87, 5774-5776, (2000).

Nachumi, B., Fudamoto, Y., Keren, A., Kojima, K. M., Larkin, M., Luke, G. M., Merrin, J., Tcernyshyov, O., Uemura, Y. J., Ichikawa, N., Goto, M., Takagi, H., Uchida, S., Crawford, M. K., McCarron, E. M., MacLaughlin, D. E. and Heffner, R. H., "Muon Spin Relaxation Study of the Stripe Phase Order in $La_{1.6-x}Nd_{0.4}Sr_xCuO_4$," Physical Review B 58, 8760-8772, (1998).

Nakagawara, O., Shimuta, T., Makino, T., Arai, S., Tabata, H. and Kawai, T., "Epitaxial Growth and Dielectric Properties of (111) Oriented $BaTiO_3/SrTiO_3$ Superlattices by Pulsed-laser Deposition," Applied Physics Letters 77, 3257-3259, (2000).

Neaton, J. B. and Rabe, K. M., "Thoery of Polarization Enhancement in Epitaxial $BaTiO_3/SrTiO_3$ Superlattices," Applied Physics Letters 82, 1586-1588, (2003).

Niedermayer, C., Bernhard, C., Blasius, T., Golnik, A., Moodenbaugh, A. R. and Budnick, J. I., "Common Phase Diagram for Antiferromagnetism in $La_{2-x}Sr_xCuO_4$ and $Y_{1-x}Ca_xBa_2Cu_3O_6$ as Seen by Muon Spin Rotation," Physical Review Letters 80, 3843-3846, (1998).

Nishiyama, Z., Martensitic Transformations, New York, Academic Press (1978).

Nix, W. D., "Mechanical Properties of Thin Films," Metallurgical Transactions A 20A, 2217, (1989).

O'Neill, D., Bowman, J. and Gregg, J. M., "Investigation into the Dielectric Behavior of Ferroelectric Superlattices Formed by Pulsed Laser Deposition," Journal of Materials Science: Materials in Electronics 11, 537-541, (2000).

144

Ong, L. H., Osman, J. and Tilley, D. R., "Dielectric Hysteresis Loops of First-order Ferroelectric Bilayers and Antiferroelectrics," Physical Review B 65, 134108, (2002).

Orihara, H., Hashimoto, S. and Ishibashi, Y., "A Theory of D-E Hysteresis Loop Based on the Avrami Model," Journal of the Physical Society of Japan 63, 1031-1035, (1994).

Orihara, H. and Ishibashi, Y., "A Statistical Theory of Nucleation and Growth in Finite Systems," Journal of the Physical Society of Japan 61, 1919-1925, (1992).

Ostenson, J. E., Bud'ko, S., Breitwisch, M. and Finnemore, D. K., "Flux Expulsion and Reversible Magnetization in the Stripe Phase Superconductor $La_{1.45}Nd_{0.40}Sr_{0.15}CuO_4$," Physical Review B 56, 2820-2823, (1997).

Otsuka, K., Wayman, C. M., Ed. Shape Memory Materials. Cambridge, Cambridge University Press, (1998).

Papageorgiou, T. P., Herrmannsdorfer, T., Dinnebier, R., Mai, T., Ernst, T., Wunschel, M. and Braun, H. F., "Magnetization Anomalies in the Superconducting State of $RuSr_2GdCu_2O_8$ and the Magnetic Study of Sr_2GdRuO_6," Physica C 377, 383-392, (2002).

Pertsev, N. A., Tagantsev, A. K. and Setter, N., "Phase Transitions and Strain-induced Ferroelectricity in $SrTiO_3$ Epitaxial Thin Films," Physical Review B 61, R825-R829, (2000).

Pintilie, L., Boerasu, I. and Gomes, M. J. M., "Simple Model of Polarization Offset of Graded Ferroelectric Structures," Journal of Applied Physics 93, 9961-9967, (2003).

Popov, Y. F., Kadomtseva, A. M., Vorob'ev, G. P., Sanina, V. A., Zvezdin, A. K. and Tehranchi, M. M., "Low-temperature Phase Transition in $EuMn_2O_5$ Induced by a Strong Magnetic Field," Physica B 284-288, 1402-1403, (2000).

Poullain, G., Bouregba, R., Vilquin, B., Rhun, G. L. and Murray, H., "Graded Ferroelectric Thin Films: Possible Origin of the Shift Along the Polarization Axis," Applied Physics Letters 81, 5015, (2002).

Qu, B. D., Evstigneev, M., Johnson, D. J. and Prince, R. H., "Dielectric Properties of $BaTiO_3/SrTiO_3$ Multilayered Thin Films Prepared by Pulsed Laser Deposition," Applied Physics Letters 72, 1394-1396, (1998).

Qu, B. D., Zhong, W. L. and Prince, R. H., "Interfacial Coupling in Ferroelectric Superlattices," Physical Review B 55, 11218-11224, (1997).

Rao, C. N. R., Raveau, B., Transition Metal Oxides: Structure, Properties, and Synthesis of Ceramic Oxides, New York, John Wiley & Sons (1998).

Rep, D. B. A. and Prins, M. W. J., "Equivalent-circuit Modeling of Ferroelectric Switching Devices," Journal of Applied Physics 85, 7923-7930, (1999).

Ricinschi, D., Harnagea, C., Papusoi, C., Mitoseriu, L., Tura, V. and Okuyama, M., "Analysis of Ferroelectric Switching in Finite Media as a Landau-type Phase Transition," Journal of Physics: Condensed Matter 10, 477-492, (1998).

Robert, G., Damjanovic, D., Setter, N. and Turik, A. V., "Preisach Modeling of Piezoelectric Nonlinearity in Ferroelectric Ceramics," Journal of Applied Physics 89, 5067-5074, (2001).

Roitburd, A. L., "Equilibrium Structure of Epitaxial Layers," Physica Status Solidi A 37, 329-339, (1976).

Roitburd, A. L., "Martensic Transformation as a Typical Phase Transformation in Solids," (1978).

Roitburd, A. L. and Kurdjumov, G. V., "The Nature of Martensitic Transformations," Materials Science and Engineering 39, 141-167, (1979).

Roytburd, A. L., "Elastic Domains and Polydomain Phases in Solids," Phase Transitions 45, 1-33, (1993).

Roytburd, A. L., "Thermodynamics of Polydomain Heterostructures. I. Effect of Macrostresses," Journal of Applied Physics 83, 228-238, (1998).

Roytburd, A. L., Alpay, S. P., Bendersky, L. A., Nagarajan, V. and Ramesh, R., "Three-domain Architecture of Stress-free Epitaxial Ferroelectric Films," Journal of Applied Physics 89, 553, (2001).

Roytburd, A. L. and Slutsker, J., "Deformation of Adaptive Materials. Part I. Constrained Deformation of Polydomain Crystals," Journal of the Mechanics and Physics of Solids 47, 2299-2329, (1999).

Roytburd, A. L. and Slutsker, J., "Coherent Phase Equilibria in a Bending Film," Acta Materialia 50, 1809-1824, (2002).

Sai, N., Meyer, B. and Vanderbilt, D., "Compositional Inversion Symmetry Breaking in Ferroelectric Perovskites," Physical Review Letters 84, 5636-5639, (2000).

Sai, N., Rabe, K. M. and Vanderbilt, D., "Theory of Structural Response to Macroscopic Electric Fields in Ferroelectric Systems," Physical Review B 66, 104108-104125, (2002).

Saito, K. and Kohn, K., "Magnetoelectric Effect and Low-temperature Phase Transitions of $TbMn_2O_5$," Journal of Physics: Condensed Matter 7, 2855-2863, (1995).

Salje, E. K. H., Phase Transitions in Ferroelastic and Co-elastic Crystals, Cambridge, Cambridge Univeristy Press (1990).

Sawyer, C. B. and Tower, C. H., "Rochelle Salt as a Dielectric," Physical Review B 35, 269, (1930).

Scaife, B. K. P., Principles of Dielectrics, Oxford, Clarendon Press (1998).

Schubring, N. W., Mantese, J. V., Micheli, A. L., Catalan, A. B. and Lopez, R. J., "Charge Pumping and Pseudopyroelectric Effect in Active Ferroelectric Relaxor-Type Films," Physical Review Letters 68, 1778-1781, (1992).

Schubring, N. W., Mantese, J. V., Micheli, A. L., Catalan, A. B., Mohammed, M. S., Naik, R. and Auner, G. W., "Graded Ferroelectrics: a New Class of Steady-State Thermal/Electrical/Mechanical Energy Interchange Devices," Integrated Ferroelectrics 24, 155-168, (1999).

Schwenk, D., Fishman, F. and Schwabi, F., "Phase Transitions and Soft Modes in Ferroelectric Supperlattices," Journal of Physics: Condensed Matter 2, 5409-5431, (1990).

Sepliarsky, M., Phillpot, S. R., Wolf, D., Stachiotti, M. G. and Migoni, R. L., "Long-ranged Ferroelectric Interactions in Pervoskite Superlattices," Physical Review B 64, (2001).

Sheikholeslami, A. and Gulak, P. G., "A Survey of Behavioral Modeling of Ferroelectric Capacitors," IEEE Transactions on Ultrasonics, Ferroelectrics, and Frequency Control 44, 917-924, (1997).

Shen, J. and Ma, Y.-Q., "Long-Range Coupling Interactions in Ferroelectric Superlattices," Physical Review B 61, 14279-14282, (2000).

Shen, J. and Ma, Y.-Q., "Long-range Coupling Interactions in Ferroelectric Sandwich Structures," Journal of Applied Physics 89, 5031-5035, (2001).

Shimuta, T., Nakagawara, O., Mkino, T., Arai, S., Tabata, H. and Kawai, T., "Enhancement of Remanent Polarization in Epitaxial $BaTiO_3/SrTiO_3$ superlattices with "asymmetric" Structure," Journal of Applied Physics 91, 2290-2294, (2002).

Shur, V. Y., Ponomarev, N. Y., Tonkacheva, N. A., Makarov, S. D., Nikolaeva, E. V., Shishkin, E. I., Suslov, L. A., Salashchenko, N. N. and Klyuenkov, E. B., "Fatigue in Epitaxial Lead Zirconate Titanate Films," Physics of the Solid State 39, 609-610, (1997).

Shur, V. Y. and Rumyantsev, E. L., "Kinetics of Ferroelectric Domain Structure During Switching: Theory and Experiment," Ferroelectrics 151, 171-180, (1994).

Shur, V. Y., Rumyantsev, E. L. and Makarov, S. D., "Geometrical Transformations of the Ferroelectric Domain Structure in Electric Field," Ferroelectrics 172, 361-372, (1995).

Shur, V. Y., Rumyantsev, E. L., Makarov, S. D. and Volegov, V. V., "How to Extract Information about Domain Kinetics in Thin Ferroelectric Films from Switching Transient Current Data," Integrated Ferroelectrics 5, 293-301, (1994).

Slonczewski, J. C. and Thomas, H., "Interaction of Elastic Strain with the Structural Transition of Strontium Titanate," Physical Review B 1, 3599-3608, (1970).

Slowak, R., Hoffmann, S., Liedtke, R. and Waser, R., "Functional Graded High-K $(Ba_{1-x}Sr_x)TiO_3$ Thin Films for Capacitor Structures with Low Temperature Coeffcient.," Integrated Ferroelectrics 24, 169, (1999).

Slutsker, J., Armetev, A. and Roytburd, A. L., "Morphological Transitions of Elastic Domain Structures in Constrained Layers," Journal of Applied Physics 91, 9049-9058, (2002).

Smit, J. and Wijn, H. P. J., Ferrites, New York, John Wiley and Sons (1959).

Specht, E. D., Christen, H.-M., Norton, D. P. and Boatner, L. A., "X-Ray Diffraction Measurement of the Effect of Layer Thickness on the Ferroelectric Transition in Epitaxial $KTaO_3/KNbO_3$ Multilayers," Physical Review Letters 80, 4317-4319, (1998).

Streetman, B. G., Solid State Electronic Devices, Englewood Cliffs, NJ, Prentice-Hall, Inc. (1980).

Strukov, B. A. and Levanyuk, A. P., Ferroelectric Phenomena in Crystals, Berlin, Spring-Verlag (1998).

Suzuki, T., Goto, T., Chiba, K., Shinoda, T., Fukase, T., Kimura, H., Yamada, K., Ohashi, M. and Yamaguchi, Y., "Observation of Modulated Magnetic Long-range Order in $La_{1.88}Sr_{0.12}CuO_4$," Physical Review B 57, R3229-R3232, (1998).

Sze, S. M., Physics of Semiconductor Devices, New York, John Wiley & Sons (1981).

Tagantsev, A. K., Pawlaczyk, C., Brooks, K. and Landivar, M., "Depletion and Depolarizing Effects in Ferroelectric Thin Films and Their Manifestations in Switching and Fatigue," Integrated Ferroelectrics 6, 309-320, (1995).

Tagantsev, A. K., Pawlaczyk, C., Brooks, K. and Setter, N., "Built-in Electric Field Assisted Nucleation and Coercive Fields in Ferroelectric Thin Films," Integrated Ferroelectrics 4, 1-12, (1994).

Tang, M., Zhang, J. H. and Hsu, T. Y. X. Z., "One-dimensional Model of Martensitic Transformations," Acta Materialia 50, 467-474, (2002).

Tranquada, J. M., Axe, J. D., Ichikawa, N., Moodenbaugh, A. R., Nakamura, Y. and Uchida, S., "Coexistence of, and Competition between, Superconductivity and Charge-Stripe Order in $La_{1.6-x}Nd_{0.4}Sr_xCuO_4$," Physical Review Letters 78, 338-341, (1997).

Tsai, M.-H., Tang, Y-H, Dey, S. K., "Co-existence of Ferroelectricity and Ferromagnetism in 1.4 nm $SrBi_2Ta_2O_{11}$ Film," Journal of Physics: Condensed Matter 15, (2003).

Tsurumi, T., Miyasou, T., Ishibashi, Y. and Ohashi, N., "Preparation and Dielectric Property of $BaTiO_3$-$SrTiO_3$ Artificially Modulated Structures," Japanese Journal of Applied Physics Part 1 37, 5104, (1998).

Van der Merve, J. H., "Crystal Interfaces. Part II. Finite Overgrowths," Journal of Applied Physics 34, 123-127, (1963).

Wadhawan, V. K., Introduction to Ferroic Materials, Amsterdam, Gordon and Breach (2000).

Wagner, D., Romanov, A. Y. and Silin, V. P., "Magnetic Properties of Inhomogeneous Ferromagnets," Journal of Experimental and Theoretical Physics 82, 945-950, (1996).

Wang, C. L. and Tilley, D. R., "Model for Simple Order-disorder Ferroelectric Superlattice," Solid State Communications 118, 333-338, (2001).

Wang, J., Neaton, J. B., Zheng, H., Nagarajan, V., Ogale, S. B., Liu, B., Viehland, D., Vaithyanathan, V., Schlom, D. G., Waghmare, U. V., Spaldin, N. A., Rabe, K. M., Wuttig, M. and Ramesh, R., "Epitaxial $BiFeO_3$ Multiferroic Thin Film Heterostructures," Science 299, 1719-1722, (2003).

Wang, R. W. and Mills, D. L., "Onset of Long-Range Order in Superlattices: Mean-Field Thoery," Physical Review B 46, 11681-11687, (1992).

Wang, X. S., Wang, C. L., Zhong, W. L. and Zhang, P. L., "Temperature Graded Ferroelectric Films Described by The Transverse Ising Model," Physics Letters A 285, 212-216, (2001).

Wang, Z. L. and Kang, Z. C., Functional and Smart Materials: Structural Evolution and Structure Analysis, New York, Plenum (1998).

Weidinger, A., Niedermayer, C., Golnik, A., Simon, R., Recknagel, E., Budnick, J. I., Chamberland, B. and Baines, C., "Observation of Magnetic Ordering in Superconducting $La_{2-x}Sr_xCuO_4$," Physical Review Letters 62, 102-105, (1989).

Whatmore, R. W., "Pyroelectric Devices and Materials," Reports on Progress in Physics 49, 1335-1386, (1986).

Whatmore, R. W., Osbond, P. C. and Shorrocks, N. M., "Ferroelectric Materials for Thermal IR Detectors," Ferroelectrics 76, 351-367, (1987).

Wu, M. K., Chen, D. Y., Chien, F. Z., Sheen, S. R., Ling, D. C., Tai, C. Y., Tseng, G. Y., Chen, D. H. and Zhang, F. C., "Anomalous Magnetic and Superconducting Properties in a Ru-based Double Perovskite," Zeitschrift Fur Physik B 102, 37-41, (1997).

Wuttig, M., Liu, L., Tsuchiya, K. and James, R. D., "Occurence of Ferromagnetic Shape Memeory Alloys (Invited)," Journal of Applied Physics 87, 4707-4711, (2000).

Yamada, K., Lee, C. H., Kurahashi, K., Wada, J., Wakimoto, S., Ueki, S., Kimura, H., Endoh, Y., Hosoya, S., Shirane, G., Birgeneau, R. J., Greven, M., Kastner, M. A. and Kim, Y. J., "Doping Dependence of the Spatially Modulated Dynamical Spin Correlations and the Superconducting-transition Temperature in $La_{2-x}Sr_xCuO_4$," Physical Review B 57, 6165-6172, (1998).

Zheng, H., Wang, J., Lofland, S. E., Ma, Z., Mohaddes-Ardabili, L., Zhao, T., Salamanca-Riba, L., Shinde, S. R., Ogale, S. B., Bai, F., Viehland, D., Jia, Y., Schlom, D. G., Wuttig, M., Roytburd, A. L. and Ramesh, R., "Multiferroic $BaTiO_3$-$CoFe_2O_4$ Nanostructures," Science 303, 661-663, (2004).

Ziman, J. M., Models of Disorder, Cambridge, Cambridge University Press (1979).

Index

150